Daniela Dieter

Temporary aquatic systems

Daniela Dieter

Temporary aquatic systems

Sediment phosphorus and leaf litter turnover

Südwestdeutscher Verlag für Hochschulschriften

Impressum / Imprint

Bibliografische Information der Deutschen Nationalbibliothek: Die Deutsche Nationalbibliothek verzeichnet diese Publikation in der Deutschen Nationalbibliografie; detaillierte bibliografische Daten sind im Internet über http://dnb.d-nb.de abrufbar.

Alle in diesem Buch genannten Marken und Produktnamen unterliegen warenzeichen-, marken- oder patentrechtlichem Schutz bzw. sind Warenzeichen oder eingetragene Warenzeichen der jeweiligen Inhaber. Die Wiedergabe von Marken, Produktnamen, Gebrauchsnamen, Handelsnamen, Warenbezeichnungen u.s.w. in diesem Werk berechtigt auch ohne besondere Kennzeichnung nicht zu der Annahme, dass solche Namen im Sinne der Warenzeichen- und Markenschutzgesetzgebung als frei zu betrachten wären und daher von jedermann benutzt werden dürften.

Bibliographic information published by the Deutsche Nationalbibliothek: The Deutsche Nationalbibliothek lists this publication in the Deutsche Nationalbibliografie; detailed bibliographic data are available in the Internet at http://dnb.d-nb.de.

Any brand names and product names mentioned in this book are subject to trademark, brand or patent protection and are trademarks or registered trademarks of their respective holders. The use of brand names, product names, common names, trade names, product descriptions etc. even without a particular marking in this works is in no way to be construed to mean that such names may be regarded as unrestricted in respect of trademark and brand protection legislation and could thus be used by anyone.

Coverbild / Cover image: www.ingimage.com

Verlag / Publisher:
Südwestdeutscher Verlag für Hochschulschriften
ist ein Imprint der / is a trademark of
OmniScriptum GmbH & Co. KG
Heinrich-Böcking-Str. 6-8, 66121 Saarbrücken, Deutschland / Germany
Email: info@svh-verlag.de

Herstellung: siehe letzte Seite /
Printed at: see last page
ISBN: 978-3-8381-3115-3

Zugl. / Approved by: Berlin, Freie Universität, Diss., 2013

Copyright © 2013 OmniScriptum GmbH & Co. KG
Alle Rechte vorbehalten. / All rights reserved. Saarbrücken 2013

Directory

Preface 1
1. Summary 3
1. Zusammenfassung 6
2. General introduction and objectives 9
   2.1. Temporary streams and lakes 12
   2.2. Objectives 20
3. Leaf litter decomposition in aquatic systems – An overview 22
   3.1. Concepts and models of leaf decomposition 22
   3.2. Leaf decomposing organisms 24
   3.3. External factors impacting on leaf decomposition rates 25
   3.4. Intrinsic factors impacting on leaf decomposition rates 27
   3.5. Leaf decomposition and water level fluctuations 29
4. Preconditioning effects of intermittent stream flow on leaf litter decomposition 31
   4.1. Abstract 31
   4.2. Introduction 32
   4.3. Materials and Methods 34
   4.4. Results 41
   4.5. Discussion 49
   4.6. Acknowledgments 55
   4.7. Appendix 56
5. Light-mediated and anoxic leaf preconditioning at intermittent stream flow affects microbial colonization and mass loss rates 58
   5.1. Abstract 58
   5.2. Introduction 59
   5.3. Methods and Material 61
   5.4. Results 70
   5.5. Discussion 78
   5.6. Acknowledgements 84
   5.7. Appendix 85
6. Phosphorus dynamics in sediments – An overview 89

	6.1.	The aquatic phosphorus cycle and phosphorus forms	89
	6.2.	Uptake and mobilization mechanisms of phosphorus in sediments	94
	6.3.	Phosphorus dynamics and water level fluctuations	98
7.	**Effects of drying on phosphorus uptake in re-flooded lake sediments**		**103**
	7.1.	Abstract	103
	7.2.	Introduction	104
	7.3.	Methods and Material	106
	7.4.	Results	109
	7.5.	Discussion	120
	7.6.	Acknowledgements	129
	7.7.	Appendix	129
8.	**Synthesis**		**130**
9.	**References**		**137**

List of Figures

Figure 2.1 Dry period in temporary streams. From upper left to lower right: a) Strong river channel contraction in the river Evrotas, Greece, August 2009. b) Beginning of stagnant pool formation in a tributary stream to the river Candelaro, Italy, September 2009. c) Totally dried streambed of the Demnitzer Mühlenfließ, Germany, May 2010. d) Installation of piezometer in the dry Demnitzer Mühlenfließ in May 2010. Photos by Daniela Dieter. 14

Figure 2.2 Water temperature and water table fluctuations relative to the sediment surface (0 m) in a temperate intermittent stream (Demnitzer Mühlenfließ, eastern Germany), determined with a piezometer in a plastic pipe deployed in the streambed (Figure 2.1d). 15

Figure 2.3 Proliferation of filamentous algae in the drying stream bed of the river Evrotas, Greece, August 2009. Photo by Daniela Dieter. 17

Figure 2.4 Sediment respiration rates during drying measured as oxygen consumption in the headspace of mesocosms filled with sediments from the river Evrotas, Greece (13% loss on ignition, grain size 100% < 630µm). Sediments were air-dried and repeatedly rewetted to 100% of water holding capacity. 19

Figure 3.1 Conceptual model for leaf litter decomposition with focus on the main processing products (modified after Gessner et al. 1999, Photo by Daniela Dieter (2010) of a *Quercus petreae* leaf decomposed to the tougher structure of veins) 23

Figure 4.1 Percentage of remaining ash free dry mass (AFDM) of control and preconditioned (irradiation with UV-VIS light, anoxic pool environment) *P. tremula* leaves in coarse mesh (8 mm) and fine mesh (0.5 mm) litter bags after 10 d (1st sampling) and approximately 45 d (2nd sampling) in five temporary streams (C: Candelaro, D: Demnitzer, F: Fuirosos, T: Taibilla, V: Vallcebre; mean ± 1SD, n = 4). 44

Figure 4.2 Concentration of ergosterol in remaining ash free dry mass (AFDM) of control and preconditioned (irradiation with UV-VIS light, anoxic pool environment), decomposing *P. tremula* leaves in fine mesh (0.5 mm) litter bags after 10 d (1st sampling) and approximately 45 d (2nd sampling) of exposure in five temporary streams (C: Candelaro, D: Demnitzer, F: Fuirosos, T: Taibilla, V: Vallcebre; mean ± 1SD, n = 4). 46

Figure 4.3 Relative abundance (individuals g-1 leaf ash free dry mass (AFDM)) of invertebrate functional feeding groups in coarse mesh bags (8 mm) associated with control and preconditioned (irradiation with UV-VIS light, anoxic pool environment) decomposing *P. tremula* leaves in five temporary streams (mean of 4 replicate pool sites) after 10 d (1st sampling, left panel) and approximately 45 d (2nd sampling, right panel). Data were log10(x+1) transformed to adjust scales. 48

Figure 5.1 Laboratory procedure for preconditioning of leaves for 21 d (storage at dark and dry conditions, irradiation with UV-VIS light, incubation in anoxic water) and in-stream immersion using coarse and fine mesh bags that were sampled at two dates (after 10 d and again after 20, 31, 38, or 40 d depending on the leaf species). DM: dry mass, LOI: loss on ignition, DGGE: denaturing gradient gel electrophoresis. 63

Figure 5.2 Proportions of chemical compounds in the leaf tissue of four species (mean ± SD, n = 4). Leaves were leached for 24 h following storage at dark and dry conditions (control) and irradiation with UV-VIS light for 21 d or incubated in anoxic water for 21 d. Asterisks indicate significant differences between treatment and the control (Student's *t*-test, *** $P < 0.001$, ** $P < 0.01$, * $P < 0.05$). DM: dry mass of leaves. 72

Figure 5.3 Concentration of ergosterol (mean ± SD, n = 4) in leaves of four species immersed in a stream for a) 10 d (1st sampling) and b) 20 to 40 d (2nd sampling) depending on the leaf species. Prior to immersion, leaves were preconditioned by irradiation with UV-VIS light and by incubation under anoxic conditions or stored dark and dry (control). DM: dry mass of leaves. 73

Figure 5.4 NMDS plots based on fungal (upper panel) and bacterial (lower panel) DGGE community profile. Ellipses calculated on a 95% confidence level are shown around their centroids for a) & d) leaf species, b) & e) preconditioning treatments, and c) & f) sampling dates for the time series of *P. tremula* leaves. Leaf tissue components that were potentially underlying the separation of leaf species were fitted into the plots a) & d). Stress values are given for each plot. 74

Figure 5.5 Abundance of macroinvertebrates assigned to selected feeding groups colonizing leaves of four species in coarse mesh litter bags deployed in a stream for a) 10 d and b) 20 to 40 d (depending on leaf species). Prior to immersion, leaves were preconditioned by irradiation with UV-VIS light and by incubation under anoxic conditions or stored dark and dry (control). DM: dry mass of leaves. 76

Figure 5.6 Bacterial and fungal communities on decomposing leaf litter in the intermittent stream Demnitzer Mühlenfließ, NE Germany. 87

Figure 5.7 Cluster analysis for fungal communities on preconditioned (c : control, uv: UV-VIS irradiated, anox: incubated in anoxic water; replicates 1,2,3) *P. tremula* leaves after 0 d, 10 d, and 38 d of decomposition in the intermittent stream Demnitzer Mühlenfließ, NE Germany. 88

Figure 6.1 Conceptual model of phosphorus (P) cycling in the sediment (grey-shaded) and overlying water column of aquatic systems. PIP: particular inorganic P, DIP: dissolved inorganic P, POP: particular organic P, DOP: dissolved organic P. 91

Figure 6.2 Relationship between the speed of an advancing water front and the increase in soluble reactive phosphorus (SRP) concentration of the stream water between a fixed point and the water front during surface flow recovery in a temperate intermittent stream (November 2009, Demnitzer Mühlenfließ, eastern Germany). 100

Figure 6.3 Cumulative release of soluble reactive phosphorus (P) from air-dried leaves of *Quercus petreae, Populus tremula, Alnus glutinosa,* and *Fraxinus excelsior*. Leaves from each species were incubated for a total of 22 days in separate water tanks containing artificial stream water (Fischer et al. 2006). 102

Figure 7.1 Rate of phosphorus (P) uptake (mean ± sdv, n = 4) in sediment columns flooded with water containing 2 mg P L^{-1}. Four columns contained original wet sediments and four columns contained previously dried sediments. 112

Figure 7.2 Changes in phosphorus (P) fractions as determined by sequential extraction of lake sediments that were re-flooded with P-enriched water (constantly restocked to 2 mg L^{-1}) over 36 weeks after having been dried or not dried (wet): a) & b) changes referring to the original wet sediment material, c) changes referring to dry sediment, isolating the effect of P addition. TP: total P, BD: bicarbonate/dithionite, SRP: soluble reactive P, NRP: non-reactive P, DM: dry mass of sediment. The dashed lines indicate the boundaries between the upper and lower layers (0 - 2.5 cm, 2.5 - 12.5 cm) and the wet base material (> 12.5 cm) that were originally sampled from the lake before homogenization, drying, and refilling of sediment columns. Values are means of laboratory duplicates obtained from pooled sediment samples. Fractions with negligible changes were excluded here (HCl-P, Residual-P). 114

Figure 7.3 Fractions of phosphorus (P, given in µg cm^{-2} per 0.5 cm depth) determined by sequential extraction of lake sediments that were re-flooded with P-enriched water (constantly restocked to 2 mg L^{-1}) over 36 weeks after having been dried or not dried (wet). TP: total P, BD: bicarbonate/dithionite, SRP: soluble reactive P, NRP: non-reactive P. Dashed lines indicate the boundaries between the upper and lower layers (0 - 2.5 cm, 2.5 - 12.5 cm) and the wet base material (> 12.5 cm) that were originally sampled from the lake before homogenization, drying, and refilling of sediment columns. Values are means of laboratory duplicates obtained from pooled sediment samples. 115

Figure 7.4 Changes in the concentrations of Fe^{2+} and Mn^{2+}, determined from a bicarbonate/dithionite (BD) extract during sequential fractionation of lake sediments that were re-flooded in sediment columns with P-enriched water over 36 weeks after having been dried or not dried (wet). Grey-shaded areas indicate changes in concentrations of Fe^{2+} and Mn^{2+}. The dashed line indicates the molar Fe:P ratio in the BD fraction at the end of re-flooding. 117

Figure 7.5 Vertical profiles for concentrations of soluble reactive phosphate (SRP), ferrous iron, and sulfate in the pore water of sediment columns over the course of 36 weeks of P addition to the overlying water column of previously dried and not dried (wet) lake sediments. 119

Figure 7.6 Simplified scheme of phosphorus (P) uptake and mobilization in lake sediments that are flooded with P-enriched water compared to sediments that are dried previous to re-flooding with P-enriched water. Shown dynamics are reduced to the most pronounced processes and changes in P pools with respect to the effect of drying. Stacked bars indicate the concentration of sequentially extractable P pools in the sediment at a particular layer. Eh: redox potential, P_{org}: organic P, MeO~P: NaOH-extractable P (mainly metal oxide-bound), P_{diss}: dissolved P. 124

Figure 7.7 Phospho-lipid fatty acid (PLFA) extracted from sediments (0 - 1 cm) of lake Müggelsee (start) and after having been dried or not dried (wet) and re-flooded with P enriched water for 36 weeks. PLFA were extracted following the procedure of Steger et al. (2011) using gas chromatography and mass spectrometry detection. 129

List of Tables

Table 4.1 Geographic characteristics of five temporary stream sites. 35

Table 4.2 Physical and chemical characteristics of five temporary stream sites during the leaf decomposition experiment (mean ± 1SD, n = 12 per stream, i.e. 3 sampling dates at each of four pool sites). 36

Table 4.3 Tissue components (mean ± 1SD, n = 4) of air-dried *P. tremula* leaves (initial) and comparison of tissue components of leaves stored under dark and dry conditions and leached for 24 h (control) with preconditioned leaves that were (i) exposed to irradiation with UV-VIS light and leached for 24 h (irradiated), and (ii) incubated and leached in anoxic water for 21 d. 42

Table 4.4 Paired Wilcoxon signed rank tests on the effect of preconditioning (irradiation with UV-VIS light, anoxic pool environment) on leaf mass loss of *P. tremula* leaves for two sampling dates (10 d and approximately 45 d) and two mesh sizes in five temporary streams. 45

Table 4.5 List of invertebrate taxa found in leaf litter bags (8 mm mesh size) from a decomposition experiment of *P. tremula* leaves in four temporary streams.. 56

Table 5.1 Tissue components (mean ± 1SD, n = 4) of air-dried senescent leaves of four deciduous tree species. AFDM: ash free dry mass. Ratios are based on percentages.. 62

Table 5.2 Values (mean ± 1SD, n = 20) of the main physical and chemical water characteristics during a leaf mass loss experiment in the temporary stream Demnitzer Mühlenfließ. DO: dissolved oxygen, DIC: dissolved inorganic carbon, DOC: dissolved organic carbon, DN: total dissolved nitrogen, SRP: soluble reactive phosphorus, DP: total dissolved phosphorus.. 65

Table 5.3 Primer systems and DGGE conditions as used for the analysis of bacterial and fungal community structure on decomposing leaf litter in a temporary stream.. 69

Table 5.4 Amount of nutrients leached from packs of air-dried leaves (mg element leached per g dry weight of leaves) during 24 h of incubation in artificial stream water at 20°C in mesocosms. Previously to leaching, control leaves were stored under dark and dry conditions and irradiated leaves were exposed to UV and daylight fluorescent lamps. SRP: soluble reactive phosphorus, DP: dissolved phosphorus, DN: dissolved nitrogren, DOC: dissolved organic carbon. 71

Table 5.5 Linear leaf mass loss rate k' (mean ± 1SD, n = 4) for 4 leaf species in coarse and fine mesh bags as calculated from the second sampling (<50% leaf mass remaining in coarse mesh bags) in the stream Demnitzer Mühlenfließ. Prior to immersion, leaves were preconditioned by irradiation with UV-VIS light and by incubation under anoxic conditions or stored dark and dry (control). 77

Table 5.6 Statistics on microbial community structure in different decomposing leaf species at different sampling dates and from different preconditioning treatments. na: not applicable, F: *Fraxinus excelsior*, Q: *Quercus petraea*. 85

Table 7.1 Concentrations of phosphorus (P) fractions as determined by the sequential extraction of original wet and laboratory dried lake sediments taken from two different layers. DM, sediment dry mass, TP: total P, SRP: soluble reactive P, NRP: non-reactive P, BD: bicarbonate/dithionite. Values are means of laboratory duplicates. 110

Table 7.2 Experimentally-determined sorption models for the uptake of phosphorus (P) in original wet and laboratory-dried lake sediments taken from two different layers. c: equilibrium P concentration in mg P L^{-1}, EPC_0: equilibrium P concentration at net zero sorption (model intercept with coordinate), q: µg adsorbed P per g sediment dry mass, q_0: intercept with ordinate, q_{max}: maximum sorption capacity in µg P g^{-1}, k: model coefficient denoting sorption affinity. Values are means of laboratory duplicates. 111

Table 7.3 Concentrations of C, N, Fe, Mn, and P in lake sediments of two different layers (original) and after P enrichment of the same sediments that had previously been dried or not dried (wet). Values are means of laboratory duplicates obtained from pooled sediment samples. 116

Preface

This book is cumulatively structured. The scientific work is presented in chapters of different studies with the chapters 4, 5, and 7 representing manuscripts that were submitted to scientific journals and were accepted for publication (chapter 4) or are currently under revision or review (chapters 5 and 7). Consequently, these chapters are equally structured including introduction, description of methods, results, and discussion. They can therefore be read independently from the other chapters. The reference of each submitted manuscript is given at the beginning of the particular chapter. The layout and formatting of the submitted manuscripts were adjusted to the layout of this book to ensure consistency. The chapters representing the different studies are supplemented by two overview chapters giving a detailed introduction to the separate topics (chapters 3 and 6). Study chapters and overview chapters form the main part, which is framed by a general introduction (chapter 2) and a general synthesis (chapter 8) providing context, current scientific knowledge, and prospects. The general synthesis does not repeat the discussions from the studies chapters but answers the purpose of placing the conclusions into a larger context. Cited references of all chapters are summarized in one References chapter at the end (chapter 9) to facilitate reading and to avoid repetition.

Preface

1. Summary

Water level fluctuations are a global phenomenon creating temporary aquatic systems. Recent trends in climate and land use changes have led to the spatio-temporal expansion, meaning that temporary streams and lakes show longer periods of low water level or that currently permanent systems switch to a temporary regime. Ecological effects of droughts have been widely studied, but the underlying ecological and physicochemical processes at the transition between dry and wet stages are poorly understood. Temporary aquatic systems are widely disregarded in science and also in management or conservation plans. Drying and re-flooding often involve the alternation of aerobic and anaerobic conditions, thus redox-sensitive processes are greatly affected. The decomposition of organic matter and nutrient dynamics are generally regarded as key ecosystem processes and are both sensitive to changing redox conditions making them valuable indicators for functional ecosystem health.

The input of organic material, such as leaf litter, may be the most important carbon source for the aquatic communities especially in streams. In temporary streams, peak leaf fall often coincides with cessation of flow so that leaves accumulate at the surface of the dry streambed or in residual stagnant pools, where they are subject to physicochemical preconditioning before subsequent decomposition in flowing water. Hardly any information existed, however, on how these preconditioning processes could affect leaf decomposition when stream flow has recovered. The experiments described in this book showed that photodegradation as occurring by solar radiation on dry streambeds and anaerobic fermentation as occurring in anoxic ponds enhanced the leaching of nutrients and labile carbon compounds. This reduced the leaf quality considering it a substrate for decomposer communities such as

Summary

macroinvertebrates and microorganisms. Effects on macroinvertebrate assemblages were not detected, but living fungal biomass was repressed and a change in the fungal community structure was indicated. As a result, leaf decomposition rates in flowing water were reduced for preconditioned leaves, which held true for a range of streams and leaf species differing in chemical characteristics and quality. The results suggest that in streams developing seasonal flow intermittence, preconditioning will influence leaf litter processing towards lower rates of microbially-mediated turnover and towards poorer quality of downstream-transported material.

The observed leaching of nutrients during preconditioning of leaves was particularly pronounced for phosphorus, which plays a key role in determining the trophic state of an aquatic system. The availability of phosphorus in the water column is controlled by the capability of the sediment to retain additional phosphorus input, which is therefore regarded as one of the most important ecosystem functions. Because phosphorus cycling is highly redox-sensitive, changes due to drying and re-flooding of the sediments may involve shifts in phosphorus uptake and mobilization. The experiments described in this book revealed that drying mobilized more stable phosphorus fractions, stimulated the mineralization of organic phosphorus compounds, and increased the proportion of labile and reductant-soluble fractions. Drying reduced the phosphorus sorption affinity and sorption capacity of the sediment, but also led to a sediment compaction, which in contrast enhanced initial phosphorus uptake rates. Following re-flooding, the compaction due to drying also induced the development of a sharp redoxcline below which P was mobilized. The results indicated that even a single drying event can result in the transformation of phosphorus components into more labile forms, which are accumulated in the near-surface sediment layer, and therefore raise the potential of pulsed P release under reducing conditions.

Summary

The results of the performed experiments provide detailed insights into the underlying processes that influence organic matter turnover and phosphorus dynamics due to drying and rewetting. Moreover, they provide important background information for the development and improvement of aquatic monitoring and management tools with respect to climate and land-use change.

1. Zusammenfassung

Wasserstandsschwankungen sind ein globales Phänomen, welches temporäre aquatische Systeme erschafft. Die gegenwärtige Entwicklung des Klima- sowie Landnutzungswandels hat bereits zu einer räumlich-zeitlichen Ausdehnung geführt, sodass temporäre Fließgewässer und Seen länger andauernde Trockenperioden niedrigen Wasserstands aufzeigen oder dass gegenwärtig perennierende Gewässer in ein temporäres Regime übergehen. Die ökologischen Auswirkungen von Dürren wurden vielfach untersucht, jedoch sind die zugrundeliegenden physikochemischen Prozesse beim Wechsel zwischen Trocken- und Nassperioden noch unzureichend aufgeklärt. Trocknung und Wiedervernässung bedeuten oft einen Wechsel zwischen aerobischen und anaerobischen Verhältnissen, sodass insbesondere redox-sensitive Prozesse durch den Wechsel beeinflusst werden. Die Zersetzung von organischem Material und die Dynamik von Nährstoffen sind Schlüsselprozesse in einem Ökosystem. Beide sind abhängig von den vorherrschenden Redoxbedingungen, was sie zu wertvollen Indikatorprozessen macht.

Der Eintrag von allochthonem organischem Material, wie beispielsweise Laubblätter, stellt die wohl wichtigste Kohlenstoff- und damit Energiequelle für aquatische Zönosen, vor allem in Bächen, dar. In temporären Bächen fallen Laubabwurf und Unterbrechung des Fließens meist zeitlich zusammen, wodurch sich die Laubstreu auf dem trockenen Bachbett oder in verbleibenden stehenden Pfützen akkumuliert. Hier sind die Blätter einer physikochemischen Präkonditionierung ausgesetzt bevor sie im wieder fließenden Gewässer weiter zersetzt werden. Es gab bisher jedoch kaum Hinweise darüber, wie diese Präkonditionierungsprozesse den weiteren Abbau der Laubstreu im wieder fließenden Gewässer beeinflussen. Die experimentellen Untersuchungen, die in der vorliegenden Dissertation dargelegt werden, zeigten, dass

Zusammenfassung

Photodegradation, wie beispielsweise durch Sonneneinstrahlung auf dem trockenen Bachbett, sowie anaerobe Fermentationsprozesse, wie beispielsweise in anoxischen Zonen der verbleibenden Pfützen, die Ausspülung von Nährstoffen und leicht verfügbaren Kohlenstoffkomponenten aus den Blättern begünstigte. Dies verminderte die Qualität des Laubs als Substrat für Zersetzergemeinschaften wie Makroevertebraten und Mikroorganismen. Effekte auf die Makroevertebratengemeinschaft konnten nicht nachgewiesen werden, jedoch war die Lebendbiomasse der Pilze vermindert und auch ihre Artenzusammensetzung zeigte Verschiebungen auf. Daraus resultierten reduzierte Abbauraten für präkonditionierte Blätter, was für eine Reihe an Bächen und Laubarten unterschiedlicher Qualitäten galt. Die Ergebnisse legen nahe, dass in Bächen, die künftig ein temporäres Regime entwickeln, die Präkonditionierung die Zersetzung organischen Materials hinsichtlich verminderter mikrobieller Abbauraten und hinsichtlich verminderter Qualität des stromabwärts transportierten Materials beeinflussen wird.

Die Auswaschung von Nährstoffen während der Laubpräkonditionierung wurde insbesondere für Phosphor beobachtet, welcher im Allgemeinen eine Schlüsselrolle bei der Kontrolle des trophischen Zustands eines Gewässers einnimmt. Die Verfügbarkeit von Phosphor in der Wassersäule ist abhängig von der Fähigkeit des Sediments zusätzlich eingetragene Phosphormengen aufzunehmen und zurückzuhalten (Retention). Sie wird daher als eine der wichtigsten Ökosystemfunktionen betrachtet. Da es in höchst redox-sensitiven Verbindungen vorkommt, können Wechsel von Trocknung und Wiedervernässung des Sediments zu einer Verschiebung der Retention und Mobilisierung führen. Die experimentellen Untersuchungen, die in dieser Arbeit beschrieben werden, ergaben, dass eine Sedimenttrocknung stabilere Phosphorverbindungen mobilisierte, die Mineralisierung organischer Phosphorverbindungen stimulierte und den Anteil labil gebundener und

Zusammenfassung

reduktiv löslicher Phosphorfraktionen erhöhte. Die Trocknung des Sediments verminderte seine Sorptionsaffinität und Sorptionskapazität für Phosphor, führte aber gleichzeitig zu einer Verdichtung, die im Gegensatz dazu die anfängliche Phosphoraufnahme begünstigte. Die Verdichtung bedingte auch die Ausbildung einer scharfen Redoxkline unterhalb derer Phosphor mobilisiert wurde. Die Ergebnisse lassen darauf schließen, dass bereits einzelne Trocknungsereignisse eine Verschiebung hinsichtlich labil gebundener Phosphorkomponenten bewirkt, die in der oberflächennahen Schicht akkumulieren und damit das Potential für eine pulsartige Phosphorfreisetzung unter reduzierenden Bedingungen erhöhen.

Die durchgeführten Experimente geben einen detaillierten Einblick mittels welcher Prozesse der Umsatz von organischem Material und Phosphor durch den Wechsel von Trocknung und Wiedervernässung beeinflusst wird. Sie liefern damit wichtiges Grundlagenwissen für die Entwicklung und Optimierung von Monitoring und Managementstrategien für Gewässer unter Berücksichtigung des zu erwartenden Klima- und Landnutzungswandels.

2. General introduction and objectives

Water is probably our most precious resource. We associate it with the beginning of life on earth and our most important commodity. We need freshwater for drinking, for watering the livestock and crops, and also for industrial purposes. In contrast to marine water, clean or unhazardous freshwater is scarce or naturally not equally available all over the world. Through human history, water scarcity has caused political conflicts and epidemics. Human activities have impacted on water resources by increased water consumption, sewage release, draining, impoundments, and other regulations of discharge. These direct effects are extended by indirect effects. Human activities strongly influenced the global climate change, which has resulted in an increase in evaporation and the occurrence of extreme weather events such as rainstorms and droughts (Meehl et al. 2007). These phenomena have naturally occurred in specific climatic zones of the world, but now spread to more temperate regions. Naturally occurring floods and droughts have created landscapes of temporary aquatic systems that do not permanently have surface water. Ponds may regularly dry out, lakes may show repeated water level fluctuations, and streams may flow intermittently. Temporary systems are typically known from arid and semi-arid zones in southern Europe, or Australia, yet they are not restricted to a specific climatic zone, but are a global phenomenon (Humphries and Baldwin 2003, Tockner et al. 2009). Glacial, nival, or pluvial regimes and phenomena in karstic systems are known (Larned et al. 2010). In fact, temporary waters were historically more common in central Europe, but changed into permanent wet or terrestrial systems due to discharge regulations and land drainage (Williams 2006). Temporary systems now re-occur due to restoration actions.

General introduction and objectives

Recent trends in climate and land use changes have led to a spatio-temporal expansion of temporary aquatic systems, meaning that (i) currently permanent systems switch to a temporary regime and that (ii) temporary systems show longer periods of low water level or shifts in the sequence of water level fluctuations. The first will basically touch temperate regions on the verge of becoming semi-arid or nival regions with less snowfall and totally thawed glaciers. The latter will concern semi-arid regions, such as the Mediterranean Basin, where temporary systems have historically existed. Reports by the *Intergovernmental Panel on Climate Change* predict an increase in temperature by up to 4.1 K in the summer and up to 24% decline in summer precipitation for the Mediterranean Basin by the next turn of the century. River runoff in the Mediterranean Basin has already declined during the 20th century due to intensified withdrawal and is prognosticated to decline further (20% less surface water availability) towards the end of the 21st century accelerated due to climate impacts (Mariotti et al. 2008). The effect is particularly strong (up to ~ 40% less surface water availability) for the Iberian and Apennine Peninsulas, the Balkans, and Turkey. Annual runoff is projected to decline by 6 - 36% within the next 50 years not only in southern Europe, but also in central and eastern Europe and the Alpine region accompanied by the extension and increase in occurrence of drought periods (Giorgi et al. 2004, Beniston et al. 2007, Bates et al. 2008). Not only surface water is diminished, but also groundwater reservoirs are affected. During the last 40 years groundwater tables in eastern Germany have already dropped 10 – 30 mm per year in 75% of the area (Germer et al. 2011).

We have grown aware of the partly very bad conditions of our freshwater ecosystems. Therefore the need for research and management has found broad interest, but we still do not understand all physical, chemical, and biological processes. Substantial research had been conducted on lacustrine ecosystems during the past century. Yet, research on riverine ecosystems was almost neglected until the second half of the 20th century. For temporary systems, however, the situation is the opposite (Bond et al. 2008). In addition, the scientific knowledge of flood impacts is much more detailed than that of drought impacts (Lake 2011).

In the year 2000, the European Union (EU) has passed the *Water Framework Directive* (WFD) (European Parliament and the Council 2000), which legally binds all member states to the evaluation of all water bodies and to design river basin management plans to achieve or maintain a good ecological and chemical status by 2015. This was extended by the *Floods Directive* (European Parliament and the Council 2007), but seasonal droughts were not a topic in either directive. That may be the reason why very soon, difficulties in the implementation of the EU-WFD arose.

What is a good hydrological status of a temporary river? What is a good ecological status of a temporary river? At which time of the year should the river be evaluated? And more importantly - What are the reference conditions? The EU-WFD was developed with the perspective of large permanently running rivers, but temporary streams and rivers were not mentioned in the EU-WFD. Reference conditions, especially for periods of drought, were missing so that temporary streams could not be evaluated. This has caused a delay in the implementation of the EU-WFD. Furthermore, the proposed measurement parameters are basically more of structural than functional description. Unfortunately, there is little to be learned from other regions of the world, since also in drought-prone Australia a lot of research has been done, but management strategies are still lacking (Bond et al. 2008). For an appropriate definition of reference conditions and to develop river management strategies for temporary waters, it is therefore vital to investigate not only the structural framework, but also the key functional mechanisms.

General introduction and objectives

The EU 7th framework project *Mediterranean Intermittent River Management* (MIRAGE, FP7-ENV-2007-1) was therefore set up with the purpose of studying the hydrology and ecology of those systems in more detail, defining reference conditions, and creating a tool box for the implementation of the EU-WFD on temporary rivers.

The chapters in this dissertation report the current state of knowledge and results from recent studies of ecological and physicochemical processes in temporary aquatic systems, which were conducted within the scope of the MIRAGE project.

The effects of drought periods and subsequent wet periods on ecosystem processes are poorly understood (Bond et al. 2008). Especially the effects on the transformation of carbon and nutrients are not well studied (Humphries and Baldwin 2003, Bond et al. 2008). The unpredictability and extent of drought events has represented a great challenge, thus the simulation of drought conditions in mesocosms under laboratory conditions may be the only useful alternative (Humphries and Baldwin 2003). In addition, many studies have focused on monitoring local systems, which has restricted general conclusions so far. More interdisciplinary, laboratory, and more large-scale or inter-site studies are therefore needed (Bond et al. 2008). The studies in this book focus on ecosystem functioning in temporary aquatic systems, in particular on leaf litter turnover and phosphorus cycling in streams and lakes. Before these topics are introduced and objectives are stated, the following section shall summarize the state of knowledge about temporary streams and lakes.

2.1. Temporary streams and lakes

The hydrological regime of a freshwater system is generally determined by characteristics of the catchment. These are climate conditions (recent and historic), catchment size, geology, topography, vegetation, and anthropogenic structures. Streams and rivers are generally stronger linked to the catchments than lakes. Flowing (lotic) and standing (lentic) waters with continuous presence of surface water even through periods of minor precipitation are perennial or permanent systems. Water bodies with disappearing of surface water are temporary systems and can be categorized by multiple classification schemes based on size, biome, or hydrology (Williams 2006). Hydrological classifications describe different regimes with increasing flooding variability and increasing duration of the drought period: (i) near-permanent – with occasional drying to which the biota is not adapted, (ii) intermittent or seasonal – with predictable periods of drying that lead to temporary surface water disappearance or cessation of surface water flow as well as strong water body contraction with disconnection of stream reaches, (iii) episodic – with long periods of drying and irregular flooding, and (iv) ephemeral – with surface water occasionally occurring for a short time (modified after Williams 2006 and Lake 2011). Droughts have therefore been often defined under hydrologic aspects, even though more definitions are known, depending on either a naturally or humanitarian point of view (Lake 2011). Although total drawdown of lakes and whole-stream drying exist, partial drying is the most common feature. While lakes experience water level fluctuations that expose areas of the littoral or may disconnect sub-basins of a lake, streams show different longitudinal drying patterns including (i) headwaters drying, (ii) mid-reach drying, and (iii) downstream drying (Lake 2003).

Drying causes an increasing contraction of the wetted area until surface flow ceases and the hydrological connectivity between stream reaches and tributaries is disrupted (von Schiller et al. 2011). A drying stream bed typically forms a mosaic of different hydrological mesohabitats (Gallart et al. 2012). Six aquatic states for mesohabitats could be defined gradually with decreasing

General introduction and objectives

discharge - cessation of flow - surface water disappearance - and hyporheic water disappearance. These include (i) overbank flow (hyperrheic), (ii) base flow (eurheic), (iii) strongly narrowed channel (oligorheic), and (iv) stagnant water in disconnected pool sections (arheic) (Figure 2.1, Lake 2003, Bond et al. 2008, Gallart et al. 2012).

Figure 2.1 Dry period in temporary streams. From upper left to lower right: a) Strong river channel contraction in the river Evrotas, Greece, August 2009. b) Beginning of stagnant pool formation in a tributary stream to the river Candelaro, Italy, September 2009. c) Totally dried streambed of the Demnitzer Mühlenfließ, Germany, May 2010. d) Installation of piezometer in the dry Demnitzer Mühlenfließ in May 2010. Photos by Daniela Dieter.

In addition, shallow pools may also dry out completely during elongated periods of drought or if the stream bed is very permeable. Subsurface flow in the hyporheic zone may persist (Figure 2.2) and connect superficially isolated pools in the (v) hyporheic state. However, also groundwater tables can drop

General introduction and objectives

and interstitial layers can dry up leaving no aquatic refuge in the (vi) edaphic state (Stanley et al. 1997, Gallart et al. 2012). The effects of droughts on groundwater can be substantial, but have been rarely studied and are poorly understood (Bond et al. 2008).

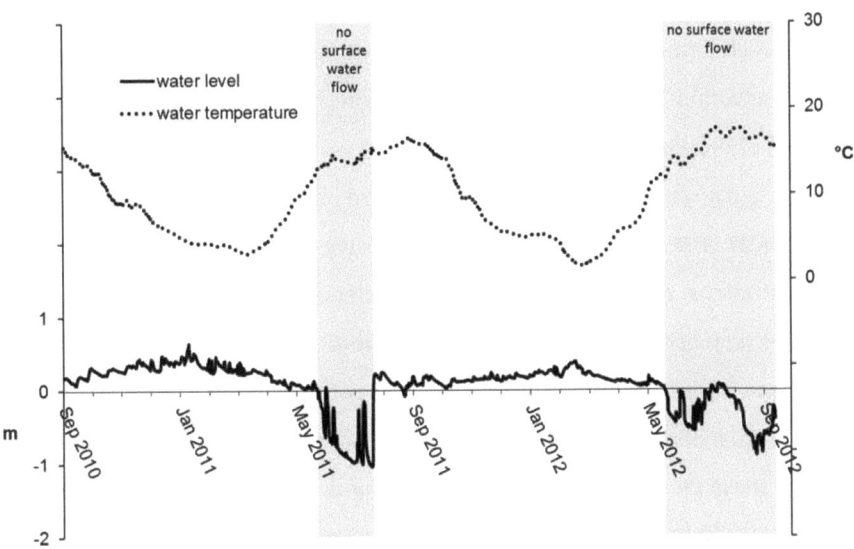

Figure 2.2 Water temperature and water table fluctuations relative to the sediment surface (0 m) in a temperate intermittent stream (Demnitzer Mühlenfließ, eastern Germany), determined with a piezometer in a plastic pipe deployed in the streambed (Figure 2.1d).

The ecological effects of droughts and fragmentation in running waters were reviewed by Lake (2003, 2011). In general, the shrinking stream size causes death of many animals by stranding. Further, populations of invertebrates and fish seeking refuge in the remaining water bodies will reach high densities. The aquatic habitats will change in physical and consequently chemical conditions. The reduction in flow velocity results in the sedimentation of transported particles such as fine sediments, particulate organic matter, and detritus. These particles may be trapped and can accumulate in stagnant pools. Also leaves from the deciduous riparian forest may be trapped or fall directly into pools and accumulate. Leaching and microbial decomposition of organic material is enhanced through higher temperatures and increases nutrient concentrations. In well sunlit pools, high primary production, and even toxic algal blooms or proliferation of filamentous algae may therefore occur (Figure 2.3, Bond et al. 2008). High primary production alternating with OM decomposition can create high fluctuations in oxygen concentration, with hypoxia likely to occur at the bottom of the water body.

The main physicochemical changes that were observed together with the loss of laminar and turbulent flow are increasing temperatures and nutrient concentrations, acidification, and oxygen depletion, especially in stagnant pool sections (von Schiller et al. 2011). In addition to pool sections in streams, the same would also apply to shallow lakes and ponds that are affected by water level fluctuations (Fernández-Aláez and Fernández-Aláez 2010, de Vicente et al. 2012). These changes may be life-threatening to groups of aquatic biota, but may favor other taxa, so that shifts in the community structure may occur. While the stream or lake contracts and gets fragmented with declining water level, the terrestrial area expands and gets connected. Terrestrial and aquatic habitats are alternating, thus benefit and harm are closely linked, making the aquatic-terrestrial transition zone a hotspot of biodiversity and process activity (flood pulse concept, Junk et al. 1989, Wantzen et al. 2008a).

Biota in naturally temporary systems seem to be very well adapted to periodical, predictable disturbances (drying and flooding) by resistance or resilience (Bond et al. 2008). They have developed different life-history strategies that include (i) avoidance (e.g. emerging insects), (ii) refuge and escape (e.g. into deep wet sediments), (iii) formation of resting stages (e.g. eggs, seeds, and other propagules), and (iv) re-dispersal (e.g. with flood wave, wind, by active flying) (Wishart 2000, Corti and Datry 2012, Steward et al.

2012). Habitats can then be re-colonized by hatching, egg deposition or return from re-connected refugia. Single species may be actually dependent on temporary waters, such as *Triops* that requires a period of desiccation before hatching upon re-flooding (Williams 2006). This affects not only aquatic but also terrestrial organisms, which together may represent a special and unique community in temporary systems (Steward et al. 2011).

Figure 2.3 Proliferation of filamentous algae in the drying stream bed of the river Evrotas, Greece, August 2009. Photo by Daniela Dieter.

Discharge usually recovers with the onset of snow melt or rain events. An abrupt onset can cause a first flood wave, while the gradual increase of discharge creates a water front progressing at various speeds. Depending on the speed, the arriving water front may scour out accumulated material and dissolved nutrients in a first flush event. Concentrations of dissolved ions such as phosphate, nitrate, ammonium, and sulfate were reported to increase rapidly (von Schiller et al. 2008). Also the speed of re-filling of lakes and ponds was reported to determine the quality and quantity of nutrient release

(Fernández-Aláez and Fernández-Aláez 2010). The water front can be used by aquatic organisms for re-dispersal and large numbers of terrestrial invertebrates drift downstream (Corti and Datry 2012). Aquatic animals may feed on accumulated dead terrestrial matter, including plant residues or drowned terrestrial animals. Aquatic beetles, for example, can surf the water front and feed on the fleeing terrestrial animals (Dieter, personal observations).

Stream metabolism is highly affected by changing water levels and is basically determined by changes in oxygen and moisture conditions. The aeration of anoxic sediments due to drying can stimulate respiratory processes and decomposition of buried organic matter. Likewise, small wetting pulses in dried sediments could stimulate respiration to a higher degree than rewetting to full saturation (McIntyre et al. 2009). However, respiration decreases again due to a lack of moisture at advanced drying stages or desiccation and is not re-enhanced until rewetting (Fromin et al. 2010). Repeated drying and rewetting was shown to stimulate the decomposition of refractory organic material (Hulthe et al. 1998). Yet, my own preliminary studies indicated that after multiple rewetting pulses and drying the respiration rates may not fully recover, most likely due to long duration of water stress and depletion of labile organic carbon substrate (Figure 2.4). However, respiration activity recovered very quickly (< 1 h) after rewetting. In general, the response of biotic and physicochemical processes on a rewetting pulse can rapidly occur within a range of < 1 min (rehydration of anhydrobiotic cyanobacteria), to < 24 h (leaching loss from dried leaf litter), and up to 14 d (emergence of adult chironomids) as reviewed by Larned et al. (2010).

The ecological effects of water level fluctuations have been less well studied in lakes than in streams and rivers (Wantzen et al. 2008b). Similarly to streams, they are dependent on the amplitude, duration, and frequency of the fluctuations as well as on the morphology of the lake (Wantzen et al. 2008a).

General introduction and objectives

Shallow lakes and lakes with a broad and shallow littoral zone will be more affected than deeper lakes due to the exposition of large sediment areas even at minor water level fluctuations (Nowlin et al. 2004). Some detailed information exists on ecological effects of water level fluctuations in reservoirs (e.g. Geraldes and Boavida 2004, Baldwin et al. 2008) and saline lakes (e.g. García and Niell 1993, Alcorlo et al. 1997). Water level fluctuations in lakes can increase the temperature variability and may lead to sediment freezing, which threatens biota living in that habitat. In addition, the relation between epilimnion and metalimnion can be markedly shifted, which could potentially lead to a change in mixis type of a lake (Naselli-Flores and Barone 2005).

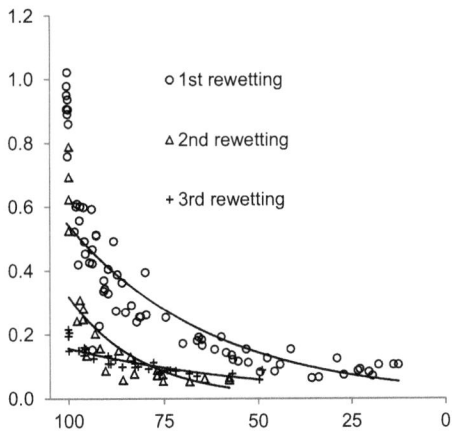

Figure 2.4 Sediment respiration rates during drying measured as oxygen consumption in the headspace of mesocosms filled with sediments from the river Evrotas, Greece (13% loss on ignition, grain size 100% < 630µm). Sediments were air-dried and repeatedly rewetted to 100% of water holding capacity.

2.2. Objectives

Drying and rewetting conditions seem to severely affect ecosystem structures and functioning in multiple ways. Hence, it is vital to assess how an increase in intermittency due to climate change and direct human impacts will affect ecosystem processes. Organic carbon turnover and nutrient cycling are generally regarded as key ecosystem processes. Hence, they represent two appropriate indicators for ecosystem changes.

Headwater streams and lake littoral zones are greatly influenced by the riparian vegetation (river continuum concept, Vannote et al. 1980). Therefore, the input of organic material, such as woody debris and especially leaf litter, plays an important role in metabolism mass balances and may be the most important carbon source for the aquatic communities. The decomposition of leaf litter from riparian groves is therefore regarded as a suitable functional indicator for metabolic activity in these systems. It is a pivotal process and an appropriate diagnostic tool to assess ecosystem health and stream integrity (Gessner and Chauvet 2002, Young et al. 2008), because it depends on physicochemical conditions as well as microbial and macroinvertebrate decomposer communities that react sensitively on habitat changes. A lot is known about leaf decomposition in lotic and lentic aquatic systems, and also on dry terrestrial sites, but little is known about leaf decomposition under alternating hydrologic conditions. The first objective of this work was therefore to assess how leaf decomposition during the dry period of temporary streams would affect further decomposition in the following period of submersion.

Nutrient turnover generally regards phosphorus and nitrogen species, since they play a key role in determining the trophic state of a system. Both are known to react sensitive on changes in oxygen supply. Especially phosphorus cycling is a highly redox-sensitive process, with changes involving shifts in the

General introduction and objectives

composition of phosphorus fractions in the sediment and impacts on the capacities of phosphorus sorption or release from the sediment. The second objective of this work was therefore to assess to what extent drying and rewetting of lake sediments would lead to shifts in phosphorus fractions and uptake capacities.

3. Leaf litter decomposition in aquatic systems – An overview

3.1. Concepts and models of leaf decomposition

Leaf breakdown after entering aquatic systems was formerly considered to consist of three successive phases (Petersen and Cummins 1974). At first, leaching rapidly releases soluble substances such as dissolved organic carbon and nutrients (~ 15% mass loss) within the first 24 h of submersion. This leaching has been reported to be necessary prior to microbial colonization, to minimize enzyme inhibiting polyphenols in the leaf tissue (Boulton 1991). In the second phase, microorganisms colonize leaf material and condition leaf tissue by decomposition. Colonization of tree leaf litter is first dominated by fungi successively followed by bacteria. This conditioning is supposed to convert leaf litter into forms that are more palatable for invertebrates, mainly shredders, who will subsequently mechanically fragment leaf litter in the third phase. After months, the soft tissue of the leaf is decomposed and only the less palatable veins have remained. However, this conceptual model has been questioned and was revised based on the experience that all phases may occur simultaneously (Gessner et al. 1999). In their concept, Gessner et al. (1999) focus on products of leaf processing, such as decomposer biomass, dissolved organic carbon, and released inorganic compounds (Figure 3.1). Qualities and quantities of these products may change gradually with advanced stages of decomposition. The stage of leaf decay is a function of time and usually referred to as the leaf decomposition rate k which determines how fast leaf mass or specific leaf tissue components are lost in a period of time.

Leaf litter decomposition in aquatic systems – An overview

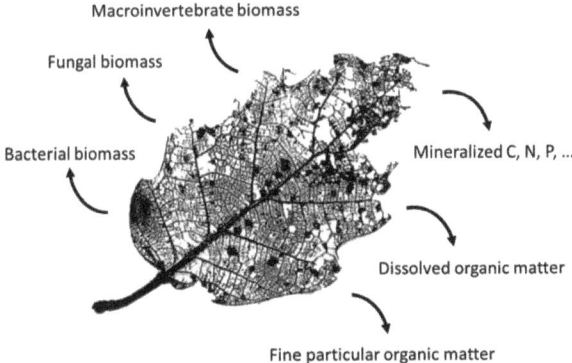

Figure 3.1 Conceptual model for leaf litter decomposition with focus on the main processing products (modified after Gessner et al. 1999, Photo by Daniela Dieter (2010) of a *Quercus petreae* leaf decomposed to the tougher structure of veins)

This breakdown function is mostly based on an exponential decay model (first-order rate equation) with *k* being the exponential decay coefficient, m_0 the initial leaf mass and m_t the remaining leaf mass after a period of time t (Bärlocher 2005):

$$m_t = m_0 \times e^{-kt} \qquad (1)$$

In addition, the first period of litter breakdown can also be assessed by a linear decay model (zero-order rate equation):

$$m_t = m_0 + k't \qquad (2)$$

The coefficients *k* and *k'* of the two models have different units and are not directly comparable with each other. The decay coefficient can be calculated from determining mass loss by measuring the percentage of remaining leaf mass after different definite time spans (Gessner and Chauvet 2002). Leaf decomposition rates greatly vary depending on many different biotic and abiotic factors, such as e.g. decomposer organisms (chapter 3.2), temperature, flow velocity, water level fluctuations (chapter 3.3), and plant species (chapter 3.4). In general, three categories of exponential daily decomposition coefficients are recognized (Petersen and Cummins 1974): fast for $k > 0.01$, medium for $0.005 \leq k \leq 0.01$ and slow for $k < 0.005$. Decomposition rates in freshwater ecosystems have been intensively studied, as reviewed in Webster and Benfield (1986). They report means of published daily decomposition coefficients k for non-woody plant material to be greatest in streams (0.0248), intermediate in lakes (0.0125), and lowest in wetlands (0.0051).

3.2. Leaf decomposing organisms

Leaf decomposition involves a range of organisms such as bacteria, fungi, macroinvertebrates, and even fishes. Hence, many different organisms can be a target to various impacts, which would ultimately affect leaf decomposition (Gessner and Chauvet 2002). Fungi are the primary colonizers and decomposers of leaf litter in aquatic systems (Gulis and Suberkropp 2006). The fungal group of aquatic hyphomycetes is well studied and believed to be the ecologically most important microbial group for leaf decomposition. Bacteria only account for usually less than 5% of the total microbial biomass on submerged decaying leaf litter (Gulis and Suberkropp 2003). Leaf decomposition rates were reported to correlate very strongly with fungal activity and fungal biomass (Gessner and Chauvet 1994). Fungi grow inside and on leaf tissues and accumulate products of decomposition and transformation. Fungal biomass can therefore account for up to 18 - 23% of total detrital mass (Gulis and Suberkropp 2006). The fungal communities seem to be sensitive to physicochemical conditions of the habitat. Baldy et al. (2002) compared diversity and biomass of aquatic hyphomycetes in different habitats and found they were higher in a flowing stream than in a floodplain pond. However, bacterial and fungal production was not significantly different among the habitats. In addition to fungi, macroinvertebrates are the second important group that contributes to leaf breakdown. Shredders not only consume leaves but also transform coarse particulate organic matter into fine particulate matter, which can be easily consumed also by other animals such as scrapers and collector-gatherers (Boulton 1991, Maamri et al. 1997, Li et al. 2009). They also discriminate among leaves to benefit from leaf-associated fungi (Graça 2001). The number of aquatic scraping or shredding invertebrates and fungal biomass is basically greater on submerged leaves than on dry sites (Boulton 1991, Langhans and Tockner 2006). Here, terrestrial invertebrates may colonize the dry sediment and feed on leaf litter and stranded aquatic biota (Steward et al. 2011).

Next to interaction and succession, the colonization and activity of leaf decomposing organisms and thus decomposition rates are sensitive to various external abiotic factors and intrinsic leaf specific factors as follows.

3.3. External factors impacting on leaf decomposition rates

Several investigations studied the abiotic factors that may affect leaf breakdown processes. These include temperature, drying and moisture content, frequency of inundation, flow velocity, oxygen concentration, radiation, nutrient concentration, acidity, as well as pollutants and biocides. Elevated temperatures affect leaf decomposition rates by stimulating microbial activity leading to enhanced production or respiration, respectively. Higher temperatures therefore lead to an increase in litter decomposition rates (Irons et al. 1994). Microbial activity also depends on a certain moisture regime, with drought severely inhibiting microbial decomposition of leaves (Baldwin and Mitchell 2000). Boulton (1991) reported lower microbial biomass on emerged leaves than on submerged leaves leading to lower decomposition rates. Litter decomposition rates have been widely reported to be higher when inundated during flooding (e.g. Boulton 1991, Ellis et al. 1999, Glazebrook and Robertson 1999, Maamri et al. 2001). Langhans and Tockner (2006) reported more than 80% of the leaf mass remaining in permanently dry conditions after 30 d of exposure, whereas only half of the leaf mass was left over after permanent inundation. The duration of inundation was also found to have a significant effect on leaf decomposition rates. Langhans and Tockner (2006) found faster breakdown rates for longer inundated leaves.

Besides temperature and moisture, biological decomposition of leaf litter is also influenced by different chemical parameters of the water body. Acidification usually entails a decline in decomposition rate caused by the inhibition of microbial and invertebrate metabolism due to the mobilization of metals (e.g. aluminum) (Webster and Benfield 1986, Dangles et al. 2004). The effect of oxygen concentration on leaf breakdown is unclear. Although low oxygen concentrations are usually regarded as slowing down decomposition rate, very inconsistent observations had been recorded in different studies (Webster and Benfield 1986). Oxygen concentrations clearly have an effect on the leaf decomposing organisms, but shifts in the community composition of microorganisms may compensate for eventual effects on decomposition rates.

Litter decomposition by microorganisms is also highly dependent on nutrient conditions of the substrate and the surrounding medium. Webster and Benfield (1986) concluded that decomposition is evidently higher in nutrient-rich systems. The effect is basically attributed to nitrogen (N), because aerobic bacteria consume N for protein production. Whereas eutrophication seems to accelerate decomposition, pollutant inputs from anthropogenic activities can remarkably inhibit litter decomposition (Gessner and Chauvet 2002, Gulis and Suberkropp 2003, Fenoglio et al. 2006, Woodward et al. 2012). Especially heavy metals from mining drainages inhibit microbial activity and subsequently decelerate leaf decomposition. Also biocides from agricultural activities in the river basin can greatly stress processing related organisms. This sensitivity for several factors makes leaf litter decomposition a good indicator (Gessner and Chauvet 2002, Young et al. 2008).

3.4. Intrinsic factors impacting on leaf decomposition rates

Last but not least, the leaves themselves are a key factor for decomposition rates. As a function of plant physiology, leaf quality is remarkably variable among species, with leaf quality encompassing (i) carbon quality, (ii) carbon to nutrient ratios, and (iii) the presence of decay inhibiting substances (Webster and Benfield 1986). Besides physically acting inhibitors as for example leaf toughness and thick waxy cuticles on conifer needles that prohibit rapid fungal invasion (Webster and Benfield 1986), most components act on physiology. The essential element found to correlate positively with faster leaf tissue breakdown is N (Webster and Benfield 1986), but also P, K, Mg, and Ca can accelerate decomposition (Zhang et al. 2008). Decay induced by microbial activity is dominated by bacteria on N limited substrates (low N:P ratio), whereas fungi dominate leaf decay on P limited substrates (high N:P ratio) (Güsewell and Gessner 2009). Besides nutrient content, the amount of labile carbon compounds determines the quality of the substrate. Woody material, such as sticks and branches, contains refractory carbon compounds, such as lignin (a phenol), which are not palatable to decomposers. The initial concentration of lignin in the leaf material is a good indicator of leaf decomposability and has repeatedly been reported as negatively correlating with decomposition rate (Webster and Benfield 1986, Gessner and Chauvet 1994, Ostrofsky 1997, Loranger et al. 2002, Sariyildiz and Anderson 2003, Zhang et al. 2008). Nevertheless, bacteria and fungi can well utilize some of the recalcitrant compounds. Aerobic heterotrophs and especially fungi use lignin more efficiently (Baldwin and Mitchell 2000). In contrast to lignin, initial holocellulose was found to correlate positively with leaf decay rates (Loranger et al. 2002). Hence, a low ratio of the proportion of holocellulose to lignocellulose would inhibit leaf litter decomposition (Bridgham and Richardson 2003). But usually, the relative concentration of the refractory carbon compounds rises during decomposition, since more labile compounds are released first. Moreover, oxidative polymerisation may cause incorporation of more simple hydrocarbons into the polyphenolic structure (Baldwin 1999). Combining the effects of lignin and nutrients, Hladyz et al. (2009) reported low lignin:N and lignin:P ratios as the principal drivers for decomposition mediated by invertebrates and microbes. This supports the results from Meentemeyer (1978), who detected a higher effect of lignin control of leaf decay rates especially in regions of higher evapotranspiration.

Besides nutrient and labile carbon content, the third important factor of leaf quality are inhibitory compounds, which suppress decomposer activity in terrestrial and aquatic ecosystems, e.g. by deterring shredders (Haase and Wantzen 2008). Polyphenols are complex hydrocarbons occurring as either hydrolysable tannins or condensed tannins (proanthocyanidins) (Hagerman 2011). These compounds can be toxic to decomposers. Further, they build complexes with polysaccharides and essential proteins like exoenzymes (Webster and Benfield 1986). Polyphenol-polysaccharide complexes build coatings on the substrate, so that a physical barrier prevents mineralization of e.g. cellulose (Kraus et al. 2003). Polyphenol-protein complexes are relatively resistant, whereas precipitation with polyphenols deactivates enzymes and decomposer feeding is therefore inhibited (Webster and Benfield 1986). Usually, negative correlations between decay rates and initial tannin concentrations were observed (Gessner and Chauvet 1994, Ostrofsky 1997, Loranger et al. 2002).

The substrate quality concerning concentrations of nutrients, carbon, and inhibiting substances in the leaf is basically a function of plant species. Yet, it can also vary among trees of the same species depending strongly on nutrient concentrations in the soil substrate and the environment (Sariyildiz and Anderson 2003). Webster and Benfield (1986) presented an overview of mean breakdown rates for woody and non-woody plant families, compiling data from studies reviewed in Petersen and Cummins (1974). Faster breakdown rates are characteristic for many non-woody families, but some grass families also show slower rates.

3.5. Leaf decomposition and water level fluctuations

Despite the broad knowledge on leaf decomposition in aquatic systems, little detailed information is available on leaf processing in temporary streams and lakes with respect to water level fluctuations or drying and rewetting, respectively. Maamri et al. (1997) and Langhans et al. (2008) reported fastest breakdown rates for lotic channel sites, intermediate rates for lentic ponds, and slowest decomposition occurred on terrestrial dry sites of an intermittent stream or a river floodplain, respectively. Hence, leaf decomposition was shown to differ among different hydrological stages. However, very few studies examined the effects of one hydrological stage preceding another during cycles of dry and wet periods. Shear stress upon rapid rewetting or due to wave impact at the littoral zone can fragment leaf litter and contribute to an observed mass loss (Pabst et al. 2008). In addition, Battle and Golladay (2001) reported highest decomposition rates under multiply flooded conditions, while frequency of inundation had no significant effect on the decomposition rate in the study by Langhans and Tockner (2006). Reduced decomposition of emersed leaves on dry streambeds were shown to accelerate after re-flooding, but did not reach decomposition rates of a comparable permanent stream section, which was most likely due to the observed inhibition of bacteria and fungi (Maamri et al. 2001). It remains, however, not fully clarified how a dry period would influence processes during the following wet period especially in intermittent streams with pronounced seasonality.

Due to water stress or autumnal timing, peak leaf fall occurs basically towards the end of the dry period (Boulton and Lake 1992, Acuña et al. 2007). As a result, most leaves accumulate either in stagnant ponds (Canhoto and Laranjeira 2007) or on dry sediments and soil (up to 80% benthic organic matter coverage of a streambed, Acuña et al. 2005, 2007). Here, the leaves are exposed to physicochemical forces that may precondition the leaves before submersion and decomposition by aquatic biota. For example, leaves can experience degradation due to leaching or desiccation and solar radiation. Brandt et al. (2009) detected CO_2 production from exposed leaves mostly due to UV radiation. Lignins in particular are susceptible to degradation caused by UVB radiation (Day et al. 2007, Austin and Ballaré 2010). Further, the destruction of leaf components due to preconditioning on the dry riverbed may severely increase the initial pulse of C, N, and P leached from leaves during inundation (Baldwin and Mitchell 2000). Moreover, leaves that are trapped in remaining ponds may leach important amounts of nutrients and labile carbon. Here, decomposition by macroinvertebrates is often restricted due to their increased mortality caused by toxic leaf leachates, oxygen depletion or acidic conditions (Canhoto and Laranjeira 2007). These conditions may additionally favor fermentative processes that reduce leaf decomposability and may have preservative effects. Overall, leaf degradation involves the destruction of refractory compounds but also the loss of labile compounds. However, it still has to be elucidated whether these contrasting effects support or inhibit leaf decomposition rates compared to non-degraded leaves upon re-flooding.

4. Preconditioning effects of intermittent stream flow on leaf litter decomposition

This chapter includes a published article, which was reprinted with kind permission of Springer Science and Business Media.

Dieter, D.; von Schiller, D.; García-Roger, E.; Sánchez-Montoya, M.; Gómez, R.; Mora-Gómez, J.; Sangiorgio, F.; Gelbrecht, J. & Tockner, K. (2011) Preconditioning effects of intermittent stream flow on leaf litter decomposition. Aquatic Sciences 73:599-609.

4.1. Abstract

Autumnal input of leaf litter is a pivotal energy source in most headwater streams. In temporary streams, however, water stress may lead to a seasonal shift in leaf abscission. Leaves accumulate at the surface of the dry streambed or in residual pools and are subject to physicochemical preconditioning before decomposition starts after flow recovery. In this study, we experimentally tested the effect of photodegradation on sunlit streambeds and anaerobic fermentation in anoxic pools on leaf decomposition during the subsequent flowing phase. To mimic field preconditioning, we exposed *Populus tremula* leaves to UV–VIS irradiation and wet-anoxic conditions in the laboratory. Subsequently, we quantified leaf mass loss of preconditioned leaves and the associated decomposer community in five low-order temporary streams using coarse and fine mesh litter bags. On average, mass loss after approximately 45 days was 4 and 7% lower when leaves were preconditioned by irradiation and anoxic conditions, respectively. We found a lower chemical quality and lower ergosterol content (a proxy for living fungal biomass) in leaves from the anoxic preconditioning, but no effects on macroinvertebrate assemblages were detected for any preconditioning treatment. Overall, results from this study

Preconditioning effects of intermittent stream flow on leaf litter decomposition

suggest a reduced processing efficiency of organic matter in temporary streams due to preconditioning during intermittence of flow leading to reduced substrate quality and repressed decomposer activity. These preconditioning effects may become more relevant in the future given the expected worldwide increase in the geographical extent of intermittent flow as a consequence of global change.

4.2. Introduction

Climate change and intensified water withdrawal for human use will dramatically increase the extent and duration of surface drying in southern Europe. In the Mediterranean Basin, an increase in summer air temperature of up to 5.5 K, a decline in precipitation by 30 - 45%, and more floods and droughts, are predicted by the end of the twenty first century (Giorgi et al. 2004). These changes are of particular consequence for temporary streams, which are found on every continent and climate, and which are the dominant freshwater type in southern Europe (Tockner et al. 2009). Intermittent flow is expected to become more common in streams across the globe due to the effects of global change (Larned et al. 2010).

Seasonal reduction in surface flow disconnects a stream from its surrounding environment - longitudinally, laterally, and vertically (Lake 2003). This fragmentation typically leads to a distinct mosaic of terrestrial and aquatic habitat types. Initially riffles and runs dry out, exposing bare dry sediments, while stagnant pools may persist longer depending on river bed morphology and the permeability of the bed sediments (Boulton and Lake 1990, Stanley et al. 1997). In such persistent stagnant pools, organic material and nutrients accumulate, and environmental conditions become harsh because of rising temperature, oxygen depletion, and acidification (Boulton and Lake 1990, Lake 2003, von Schiller et al. 2011).

Preconditioning effects of intermittent stream flow on leaf litter decomposition

In continuously flowing, forested, headwater streams, the allochthonous input of terrestrial leaves serves as a pivotal energy source (Vannote et al. 1980). However, in temporary streams, water stress may shift the timing of leaf abscission, so that leaf fall coincides with ceased flow (Boulton and Lake 1992). As a consequence, the leaves accumulate in residual pools and/or on the surface of the dry streambed, and are only transported downstream upon the first flush events (Acuña et al. 2007). Hence, in temporary streams, organic matter dynamics are controlled through the seasonal alternation of terrestrial and aquatic phases.

Organic matter decomposition in temporary streams has been studied by, e.g. Boulton and Boon (1991), Maamri et al. (1997), and Datry et al. (2011), and in floodplains and wetlands by, e.g., Glazebrook and Robertson (1999), Battle and Golladay (2001), and Langhans et al. (2008). Decomposition was usually slower for non-submersed than for submersed leaves, due to higher decomposer activity in the presence of water. In addition, decomposition was slower in standing pools than in flowing channels, likely because leaching of leaf litter results in more toxic substances and more acidic and oxygen-depleted water, which in turn reduces the numbers and activity of decomposers such as aquatic hyphomycetes and macroinvertebrates (Lake 2003, Canhoto and Laranjeira 2007, Schlief and Mutz 2007).

Limited information is available on the effect of the preconditioning that takes place during the dry phase on leaf decomposition during the subsequent wet phase. A lack of moisture on dry sediment may result in intensive desiccation of leaves, enhancing subsequent leaching losses during the first water contact (Leopold et al. 1981, Gessner et al. 1999). Furthermore, leaf abscission due to water stress in the riparian vegetation reduces canopy cover, thus leaves remaining on the sediment surface are also exposed to high solar radiation, possibly leading to photodegradation of the leaf tissue (Austin and

Vivanco 2006, Day et al. 2007, Henry et al. 2008). In contrast, leaves entering oxygen-depleted stagnant pools may be subject to leaching and fermentation during anaerobic degradation (Küsel and Drake 1996, Reith et al. 2002). Thus, leaves that enter the stream during cessation of flow might be preconditioned, prior to their further decomposition after recovery of stream flow. This preconditioning may change the chemical composition of the leaves. The latter is known to control their suitability as substrate for decomposers such as invertebrates and aquatic hyphomycetes, thereby affecting leaf decomposition rates (Leroy and Marks 2006, Yoshimura et al. 2008).

The aim of our present study was to examine how preconditioning, by photodegradation on sunlit streambeds and by anaerobic fermentation in anoxic pools, affects leaf decomposition during the subsequent flowing phase. We hypothesized that leaf preconditioning would alter leaf decomposition rates by modifying leaf chemical composition and thereby their suitability as substrate for leaf decomposers. Furthermore, we hypothesized that the effect of leaf preconditioning would be distinct for microbial and macroinvertebrate colonization of leaves because of the different sensitivity of both decomposer groups to changes in leaf chemical composition. Based on these hypotheses, we predicted differences between preconditioned and unconditioned leaves in leaf chemistry, decomposer communities, and decomposition rates mediated by microorganisms alone and together with macroinvertebrates.

4.3. Materials and Methods

4.3.1. Study sites

Field experiments were carried out in four Mediterranean temporary streams (Candelaro, Fuirosos, Taibilla, Vallcebre) and in one temperate (Demnitzer Mühlenfließ) temporary stream, which differed in catchment lithology, riparian vegetation cover, and physicochemical characteristics (Table 4.1, Table 4.2).

Preconditioning effects of intermittent stream flow on leaf litter decomposition

Table 4.1 Geographic characteristics of five temporary stream sites.

Catchment	Geographic location	Order	Elevation at study site (m a.s.l.)	Mean annual precipitation (mm)	Mean air temperature (°C)	Lithology	Riparian vegetation
Candelaro	41° 46' N, 15° 18' E	2nd	100	600	15.3	Dolomitic	Grass: *Phragmites australis*, Cyperaceae
Demnitzer	52° 21' N, 14° 11' E	3rd	40	480	9.0	Glacial sediments	Forest: *Quercus robur, Alnus glutinosa, Carpinus betulus, Fagus sylvatica*
Fuirosos	41° 42' N, 2° 34' E	3rd	150	750	16.5	Granitic	Forest: *Alnus glutinosa, Corylus avellana, Populus nigra, Platanus acerifolia*
Taibilla	38° 08' N, 2° 13' W	2nd	1200	583	13.3	Limestone, sandstone	Grass, shrub: *Scirpus holoschoneus, Salix sp., Populus alba*
Vallcebre	42° 12' N, 1° 49' E	1st	1200	924	7.3	Limestone	Grass, forest: *Populus nigra, Corylus avellana, Prunus spinosa, Pinus sylvestris, Quercus pubescens*

These diverse sites were chosen in order to replicate field experiments in a variety of temporary streams with different environmental conditions. The Candelaro (study site at a headwater stream reach) is located in the Gargano promontory in Puglia, Italy, and carries high nutrient loads as a consequence of intensive fertilizer use and groundwater use for irrigation. The Fuirosos drains the Montnegre-Corredor mountain range in Catalonia, Spain, and exhibits near-natural conditions. The Taibilla (Rambla de la Rogativa study site) lies at the eastern margin of the Baetic ranges in Murcia, Spain, and drains a highly erosive landscape, leading to high sediment and solute loads. The Vallcebre (Can Vila study site) lies at the southern margin of the Pyrenees in Catalonia, Spain, and also drains a highly erosive landscape. The Demnitzer Mühlenfließ

Preconditioning effects of intermittent stream flow on leaf litter decomposition

(Demnitzer) is a temperate lowland stream in Brandenburg, Germany, with elevated nutrient loads during high flow due to upstream agricultural sites.

Table 4.2 Physical and chemical characteristics of five temporary stream sites during the leaf decomposition experiment (mean ± 1SD, n = 12 per stream, i.e. 3 sampling dates at each of four pool sites).

Parameter	Candelaro	Demnitzer	Fuirosos	Taibilla	Vallcebre
Water temperature (°C)[a]	12.0 ± 1.7	3.9 ± 3.2	11.8 ± 5.3	11.5 ± 1.4	5.6 ± 1.5
Current velocity (cm s^{-1})	40 ± 0	20 ± 9	22 ± 16	35 ± 20	13 ± 29
Depth (cm)	17 ± 2	33 ± 10	nd	20 ± 5	29 ± 6
Specific conductance (µS cm^{-1})	997 ± 49	903 ± 32	158 ± 23	759 ± 16	1554 ± 106
pH	7.0 ± 0.4	8.2 ± 0.1	7.6 ± 0.1	8.6 ± 0.2	7.3 ± 0.9
O_2 (mg L^{-1})	10.9 ± 1.2	11.2 ± 1.3	11.2 ± 1.3	9.2 ± 1.5	12.7 ± 3.6
DOC (mg C L^{-1})	5.0 ± 1.0	13.5 ± 0.8	4.2 ± 1.1	4.0 ± 1.2	4.3 ± 1.0
NO_3^- (mg N L^{-1})	7.0 ± 0.7	9.4 ± 1.7	0.6 ± 0.5	1.2 ± 0.1	0.4 ± 0.3
SRP (µg P L^{-1})	<5	47 ± 3	8 ± 4	29 ± 1	41 ± 16

nd: not determined
[a] mean of data recorded at hourly intervals

4.3.2. Leaf preconditioning

For our model species of leaf, we chose the European aspen (*Populus tremula*). Populus species are common in riparian forests throughout Europe. However, the European aspen was absent from all study sites, thereby avoiding adaptation effects of aquatic decomposer species.

Freshly fallen senescent leaves of European aspen were collected with nets from beneath a stand of trees in a forest (52° 27' N, 13° 40' E) close to Berlin, Germany, in autumn 2009. The leaves were air-dried and stored in the dark until used in the preconditioning experiments, which were performed in winter 2009/2010 at the Institute for Freshwater Ecology and Inland Fisheries in Berlin, Germany.

Preconditioning effects of intermittent stream flow on leaf litter decomposition

We mimicked conditions in the residual pools that occur during the dry season by incubating a set of 500 g of European aspen leaves in a water tank under anoxic (< 0.1 mg DO L^{-1}) and acidic (pH 5.0) conditions at room temperature (20°C). We used artificial stream water (180 L, prepared following Fischer et al. 2006) that was inoculated with 1 L of lake water to introduce aquatic microorganisms. After 21 days, the leaves were removed from the water tank and immediately air-dried using a fan (48 h). Two subsets of four replicate mesh bags, each bag containing approximately 4.0 g of leaves, were also submersed in the water tank and sampled to determine mass loss after 24 h (initial leaching) and 21 days (full leaching) of exposure, respectively.

Photodegradation of a second set of leaves was mimicked by irradiation for 12 h per day over a period of 21 days with a combination of UV and daylight fluorescent lamps (covering wavelengths from 300 to 700 nm; UV: Cosmedico Arimed B6 [with 31% UVB of total UV], daylight: Osram Biolux 965; 50 W m^{-2} total radiation, 17 W m^{-2} UV radiation; measured with LiCor 1800 spectroradiometer). A control set of leaves was stored in the dark under dry conditions.

For the irradiation treatment and for the control, two subsets of 4 replicate mesh bags each were used to determine mass loss during the irradiation and storage in the dark for 21 days. Another two subsets were used to determine mean leaching loss by submersion in artificial stream water for 24 h under laboratory conditions, with a leaf mass to water volume ratio equal to that of the anoxic pretreatment.

To characterize the effect of preconditioning on the chemical composition of leaves, four leached leaf subsamples from the two treatments and the control were oven-dried (40°C, 48 h), milled to a fine powder, and analyzed for five parameters: (i) percentage of organic compounds (loss on ignition at 450°C), (ii) total carbon and (iii) nitrogen content (Elementar vario EL C/N elemental analyzer, Hanau, Germany), (iv) fiber contents (gravimetric lignin and cellulose determination, Gessner 2005b), and (v) phenolics (detection with a spectrophotometer following extraction with 70% acetone, Bärlocher and Graça 2005).

4.3.3. In-stream leaf decomposition

For the field experiments, fine (0.5 mm mesh size) and coarse mesh (8 mm) nylon bags were used to differentiate between microbial- and invertebrate-driven decomposition, respectively (Bärlocher 2005). Each bag was filled with approximately 4.0 g dry mass (DM) of either control leaves or preconditioned leaves (anoxic or irradiated). Field decomposition experiments were conducted concomitantly at all 5 study sites in March/April 2010 when stable flow conditions were re-established and the aquatic communities had recovered from flow cessation at all study sites. Bags containing leaves from the different treatments (control, irradiated, anoxic) were tied to an iron rod anchored in the streambed. Four pool sites per stream, located within a few hundred meters, gave quadruplicate samples for each treatment.

Field samples were collected at two time points to determine if the hypothesized effect of leaf preconditioning persisted at early and late stages of decomposition. A first set of bags was retrieved from the streams after 10 days of exposure, and a second set was retrieved after approximately 45 days, depending on the observed progress of leaf decomposition in the particular stream (38 days in Demnitzer, 40 days in Candelaro, 47 days in Vallcebre, 48 days in Fuirosos, 50 days in Taibilla). Bags were cut from the iron rods, and stored in individual plastic bags on ice in a cooler during transport. A total of 240 bags were used (5 stream sites, 4 pool sites, 3 treatments, 2 mesh sizes, 2 sampling dates), only 1 of which was lost (a coarse mesh bag from Demnitzer).

Preconditioning effects of intermittent stream flow on leaf litter decomposition

Hand held sensors were used, adjacent to the iron rods, to measure dissolved oxygen (mg DO L^{-1}), pH, specific conductance (µS cm^{-1}), and current velocity (cm s^{-1}). Measurements were made at each of the four pool sites, at three dates (date of leaf bag exposure, first sampling date, second sampling date), in each of the 5 streams (therefore n = 12 per stream). In addition, surface water was collected, filtered (0.7 µm glass fiber filters, Whatman GF/F), and stored in polyethylene bottles until analysis for dissolved nutrients. One iButton temperature logger (iBCod 22L, Alpha Mach Inc., Mont St-Hilaire, Canada) was fixed to an iron rod at each of the four pool sites to record the temperature at hourly intervals.

4.3.4. Laboratory analyses of field samples

Immediately upon arrival at the local laboratory, leaves were removed from the bags and carefully rinsed with tap water to remove any adhering debris. The remaining slurry from the coarse mesh bags was passed through a 500 µm mesh screen to retain macroinvertebrates, which were preserved in 70% ethanol until the individuals were identified, counted, and assigned to functional feeding groups (Schmidt-Kloiber et al. 2006). Invertebrate abundance was expressed as number of individuals (Ind) per g leaf ash free dry mass (AFDM).

Two discs were cut out from each of 5 randomly selected leaves from every bag (10 discs per bag) using a cork borer (11 mm diameter). One set of 5 discs was placed in an individual small polyethylene bag and sent to the IGB (kept frozen at -20°C) for analysis of ergosterol content (see below). The second set of 5 discs was placed into an aluminium pan, oven-dried (40°C, 48 h), and weighed to the nearest 0.0001 g to obtain DM. This set was subsequently ignited at 450°C in a muffle furnace to obtain the percentage of organic compounds.

Following quick drying using a fan, the bulk of leaves from each bag was oven-dried (40°C, 48 h) before weighing to the nearest 0.01 g DM. Total DM of leaves per bag was then calculated as the sum of DM of the bulk of leaves and the sets of leaf discs. The total AFDM remaining was obtained by multiplying the total DM with the percentage of organic compounds determined with the set of 5 discs. To determine the mass loss due to handling of bags, extra control leaf bags were carried to the field and back, and remaining mass was defined as 100% AFDM remaining (Benfield 2006). In addition, the initial weights of the control and irradiation samples were corrected for leaching losses occurring within the first 24 h of submersion, since leaves from the anoxic pretreatment had already been leached during 21 days of incubation in the water tank prior to initial weight determination for the field experiment. Each correction factor was determined from the mean leaching loss during 24 h as explained earlier.

The total content of ergosterol (a proxy for fungal biomass) in decomposing leaves was measured from a freeze-dried and weighed (DM) set of 5 leaf discs from each bag (see above), by microwave-assisted liquid phase extraction (Young 1995) and detection with HPLC and UV spectrometry (Gessner 2005a; Dionex UltiMate 3000 LC, Sunnyvale CA, USA). The ergosterol concentration in leaves was expressed in $\mu g\ g^{-1}$ AFDM of leaves.

At each participating laboratory, filtered water samples from the field were analyzed for the concentration of soluble reactive phosphorus (SRP), nitrate (NO_3^-), and dissolved organic carbon (DOC) using standard colorimetric methods and a TOC analyzer.

4.3.5. Data analysis

The tissue constituents of preconditioned (i.e. incubated in anoxic water tank or UV-VIS irradiated and leached) versus leached control leaves were compared using the Wilcoxon rank sum test.

Preconditioning effects of intermittent stream flow on leaf litter decomposition

The overall effects of stream (n = 5) and treatment (control, irradiation, anoxic; n = 3) on % AFDM remaining, ergosterol content, and invertebrate abundance, were tested using the Kruskal-Wallis rank sum test. Subsequently, differences in % AFDM remaining, ergosterol content, and invertebrate abundance between the two sampling dates, mesh sizes (coarse and fine), and between the treatments and the control, were estimated using the Wilcoxon signed rank test. Samples were therefore paired so that they matched in all factors (stream, sampling date, mesh size, treatment, and replicate pool site) except for the factor tested. Relations among variables were assessed using Spearman rank correlations.

Non-parametric tests were chosen throughout, because not all data met normality requirements, even after data transformations. All statistical analyses were performed using *R* statistical software (version 2.12.0, Ihaka and Gentleman 1996, http://www.r-project.org) with a significance level set at α = 0.05 for all tests.

4.4. Results

4.4.1. Environmental characteristics of the study streams

The five streams differed substantially in water quality (Table 4.2), reflecting the different land use types in their respective catchments. The temperate Demnitzer was the coldest and most nutrient-rich stream, while Fuirosos had the lowest nutrient content. Specific conductance differed by one order of magnitude among streams (158 - 1554 µS cm^{-1}). Current velocity showed high within-stream variability and peaked in Taibilla at 85 cm s^{-1}. No flood or dry events occurred during the field experiments.

Preconditioning effects of intermittent stream flow on leaf litter decomposition

4.4.2. Leaf preconditioning

Leaves incubated under anoxic conditions lost 18.5 ± 1.2% AFDM (mean ± 1SD, n = 4) during the first 24 h, and an additional 7.8 ± 1.5% AFDM after 21 days (the end of the experimental exposure). Leaves preconditioned by UV-VIS irradiation lost 3.7 ± 1.1% AFDM after 21 days, and an additional amount of 19.2 ± 1.1% AFDM during subsequent leaching (24 h exposure). In comparison, control leaves lost only 1.9 ± 0.6% AFDM after 21 days (stored dry and dark), and subsequent leaching led to an additional loss of 16.8 ± 0.7% AFDM.

Incubation under anoxic conditions led to an increase in relative fiber content and a decrease in the C:N ratio and phenolic content (Table 4.3). In contrast, in the irradiated leaves, these components did not differ significantly from the controls after leaching (Table 4.3).

Table 4.3 Tissue components (mean ± 1SD, n = 4) of air-dried *P. tremula* leaves (initial) and comparison of tissue components of leaves stored under dark and dry conditions and leached for 24 h (control) with preconditioned leaves that were (i) exposed to irradiation with UV-VIS light and leached for 24 h (irradiated), and (ii) incubated and leached in anoxic water for 21 d.

	Initial	control	irradiated	anoxic
AFDM (%)	89.8 ± 0.4	91.9 ± 0.3	90.1 ± 0.8*	88.8 ± 3.1
Lignin (%)	14.5 ± 0.6	18.1 ± 0.5	17.4 ± 0.7	21.0 ± 1.0*
Cellulose (%)	24.1 ± 0.6	30.4 ± 0.6	30.2 ± 0.2	32.6 ± 0.8*
Phenolics (%)	3.2 ± 0.3	1.5 ± 0.1	1.6 ± 0.3	1.3 ± 0.1*
C:N ratio	61.3 ± 2.6	58.3 ± 2.1	56.3 ± 2.4	51.4 ± 2.2*
Lignin:N ratio	19.0 ± 0.6	22.3 ± 0.8	20.7 ± 1.5	22.7 ± 0.6

AFDM: ash free dry mass.
Asterisks indicate a significant effect of preconditioning treatment (irradiation, anoxic pool environment) compared to control (Wilcoxon rank sum test, * $P<0.05$).

4.4.3. In-stream leaf decomposition

The % AFDM remaining after the decomposition experiments varied significantly between the five streams (Kruskal-Wallis, χ^2 = 30.94, $P < 0.001$, n =

Preconditioning effects of intermittent stream flow on leaf litter decomposition

239) (Figure 4.1). The mean % AFDM remaining at the second sampling date was 33% lower than at the first sampling date (Wilcoxon signed rank, $W = 6896$, $P < 0.001$, $n = 119$). The mean % AFDM remaining was 11% lower in coarse than in fine mesh bags (Wilcoxon signed rank, $W = 5969$, $P < 0.001$, $n = 119$). However, a few fine mesh bags contained lower % AFDM than did the corresponding coarse mesh bag (regardless of stream type, treatment, sampling date, or site). Remarkably, some fine mesh bags from Demnitzer and Taibilla in the first sampling and from Vallcebre in the second sampling had an increase in % AFDM remaining, instead of a decrease.

Overall, anoxic incubation and irradiation inhibited loss of leaf mass (Figure 4.1, Table 4.4). For the anoxic treatment, the inhibition was significant for both sampling dates and mesh sizes, with 7% more AFDM remaining on average. For the irradiation treatment, the inhibition was only significant for fine mesh bags (Table 4.4), with 4% more AFDM remaining on average. Significant preconditioning effects were detected for all study streams except for Vallcebre.

The concentration of ergosterol (a proxy for fungal biomass) in decomposing leaves differed significantly among the five streams (Kruskal-Wallis, first sampling: $\chi^2 = 13.64$, $P < 0.01$; second sampling: $\chi^2 = 27.48$, $P < 0.001$, $n = 60$). In addition, ergosterol concentrations increased significantly between the two sampling dates (Wilcoxon signed rank, $W = 0$, $P < 0.001$, $n = 60$). Mean concentrations were 51 ± 18 µg g^{-1} AFDM (mean ± 1SD, $n = 60$) at the first sampling and 160 ± 53 µg g^{-1} AFDM at the second sampling (Figure 4.2).

Preconditioning effects of intermittent stream flow on leaf litter decomposition

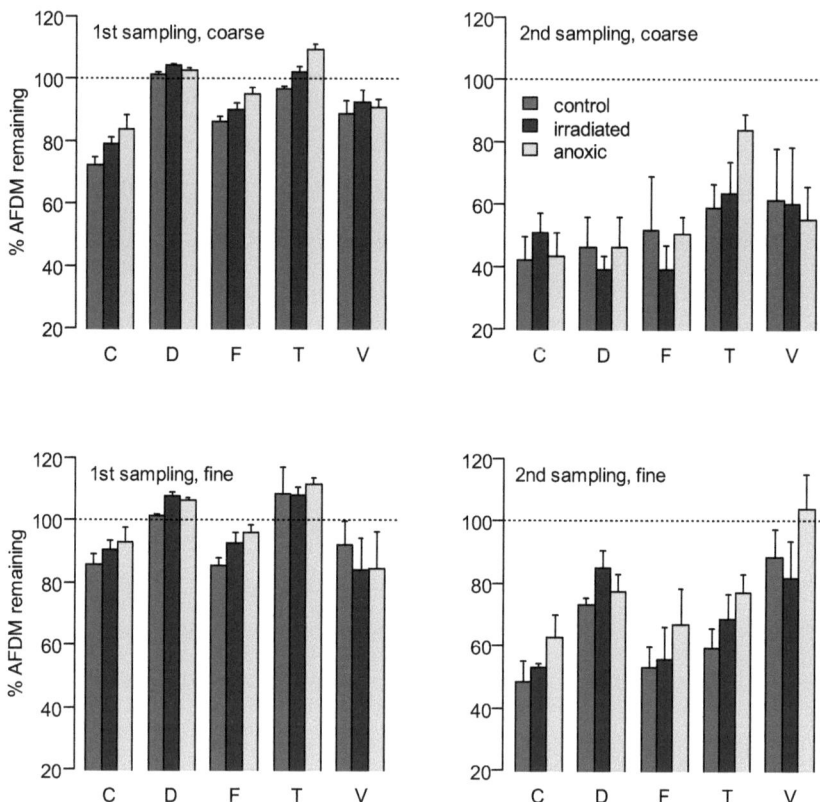

Figure 4.1 Percentage of remaining ash free dry mass (AFDM) of control and preconditioned (irradiation with UV-VIS light, anoxic pool environment) *P. tremula* leaves in coarse mesh (8 mm) and fine mesh (0.5 mm) litter bags after 10 d (1st sampling) and approximately 45 d (2nd sampling) in five temporary streams (C: Candelaro, D: Demnitzer, F: Fuirosos, T: Taibilla, V: Vallcebre; mean ± 1SD, n = 4).

Table 4.4 Paired Wilcoxon signed rank tests on the effect of preconditioning (irradiation with UV-VIS light, anoxic pool environment) on leaf mass loss of *P. tremula* leaves for two sampling dates (10 d and approximately 45 d) and two mesh sizes in five temporary streams.

	Total (n=80)	1st sampling (n=40)	2nd sampling (n=40)	coarse mesh (n=40)	fine mesh (n=40)
control vs. irradiated	2154 ** [a]	626 **	476 [a]	506 [a]	587 *
1st sampling (n=20)				174	195
2nd sampling (n=20)				91 [a]	110 *
control vs. anoxic	2557 ***	688 ***	618 **	615 **	679 ***
1st sampling (n=20)				187	227 *
2nd sampling (n=20)				120 **	128 **
anoxic vs. irradiated	2263 *** [a]	574 *	188 ** [a]	541 * [a]	605 **
1st sampling (n=20)				112	198
2nd sampling (n=20)				135 *** [a]	115 *

P-values: < 0.001 ***, < 0.01 **, < 0.05 *
[a] due to one missing value, n is reduced by 1

4.4.4. Leaf-associated fungal biomass

At the first sampling, there were no significant treatment effects on the ergosterol concentration. At the second sampling, the ergosterol concentration in leaves from the anoxic preconditioning was on average 22 µg g^{-1} AFDM lower than in the control leaves and 33 µg g^{-1} AFDM lower than in the irradiated leaves (Wilcoxon signed rank, control: $W = 159$, $P < 0.01$; irradiated: $W = 175$, $P < 0.001$, $n = 20$).

Lower concentrations of ergosterol generally corresponded to higher percentages of AFDM remaining (Spearman rank correlation, $\rho = -0.56$, $P < 0.001$, $n = 60$). In particular, the difference in the ergosterol concentration between the anoxic and control leaves, and between the anoxic and irradiated leaves, corresponded to the difference in % AFDM remaining (Spearman rank

Preconditioning effects of intermittent stream flow on leaf litter decomposition

correlation, control: ρ = -0.57, $P < 0.001$; irradiated: ρ = -0.36, $P < 0.05$, n = 20), i.e. less ergosterol was associated with slower leaf mass loss for the anoxic preconditioning.

4.4.5. Leaf-associated macroinvertebrates

The invertebrate community associated with the leaf packs consisted of 36 families (95% of all individuals were Insecta, for taxa list see Table 4.5 in chapter 4.7 Appendix). Abundances ranged from 0 Ind g^{-1} AFDM in the Demnitzer to 2,870 Ind g^{-1} AFDM in Fuirosos (Figure 4.3). Mean values at the second sampling were 66 ± 43 Ind g^{-1} AFDM (mean ± 1SD, n = 12) in Candelaro, 43 ± 35 Ind g^{-1} AFDM in Demnitzer, 1,069 ± 969 Ind g^{-1} AFDM in Fuirosos, 99 ± 68 Ind g^{-1} AFDM in Taibilla, and 28 ± 17 Ind g^{-1} AFDM in Vallcebre.

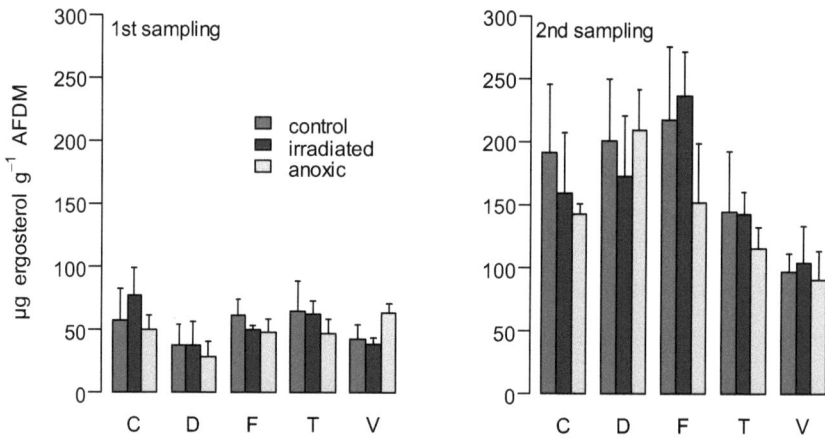

Figure 4.2 Concentration of ergosterol in remaining ash free dry mass (AFDM) of control and preconditioned (irradiation with UV-VIS light, anoxic pool environment), decomposing *P. tremula* leaves in fine mesh (0.5 mm) litter bags after 10 d (1st sampling) and approximately 45 d (2nd sampling) of exposure in five temporary streams (C: Candelaro, D: Demnitzer, F: Fuirosos, T: Taibilla, V: Vallcebre; mean ± 1SD, n = 4).

Preconditioning effects of intermittent stream flow on leaf litter decomposition

The total abundance of invertebrates and the abundance of each of the three feeding groups (shredders, grazers, collectors) were significantly different among the five streams (Kruskal-Wallis, total abundance χ^2 = 38.53, shredders χ^2 = 57.44, grazers χ^2 = 58.01, collectors χ^2 = 45.06, $P < 0.001$ for all, n = 119). Shredders were only dominant in Demnitzer (up to 58 Ind g^{-1} AFDM or up to 69% of all individuals) and were rare in the other streams, where collectors dominated (38% of all individuals on average).

Abundances of each of the three feeding groups were significantly higher at the second sampling date than at the first sampling date, except in Candelaro (significantly lower at the second date) (Wilcoxon signed rank, total abundance W = 208, shredders W = 33, grazers W = 108, collectors W = 278, $P < 0.001$ for all, n = 59). Preconditioning of leaves did not significantly affect the total invertebrate abundance, or feeding group abundance (Kruskal-Wallis, χ^2 = 0.68, $P < 0.05$ for all, n = 119).

The mean contribution of macroinvertebrates to total mass loss (based on the difference in AFDM lost between coarse and fine mesh bags) was 32%. The percentage of invertebrate contribution to mass loss was not significantly affected by preconditioning treatment (Kruskal-Wallis, χ^2 = 1.14, df = 2, $P < 0.05$, n = 119).

Preconditioning effects of intermittent stream flow on leaf litter decomposition

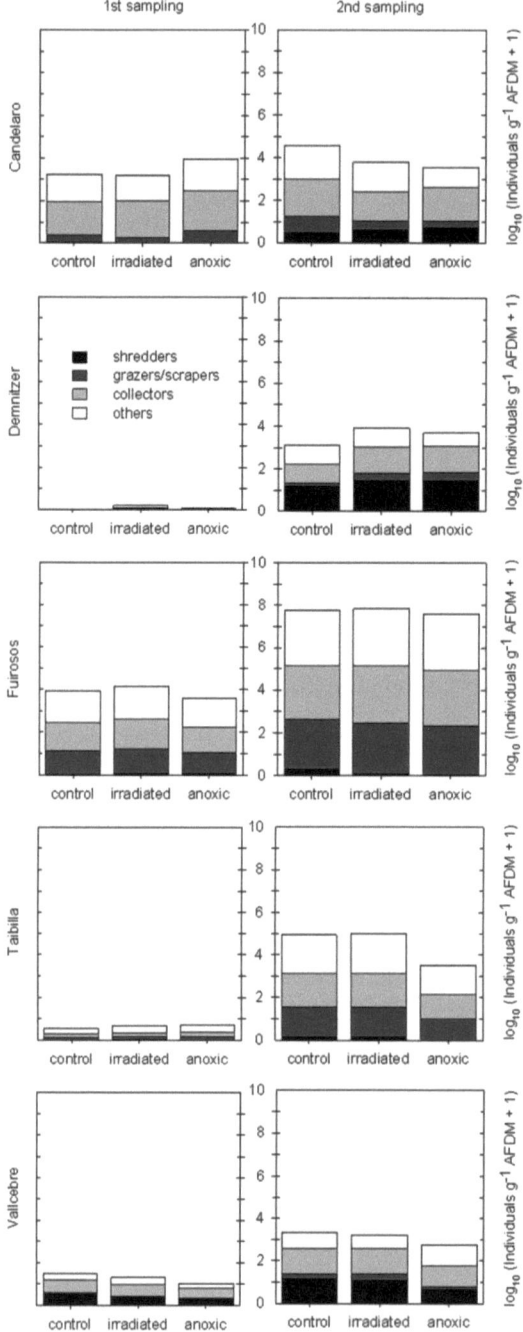

Figure 4.3 Relative abundance (individuals g-1 leaf ash free dry mass (AFDM)) of invertebrate functional feeding groups in coarse mesh bags (8 mm) associated with control and preconditioned (irradiation with UV-VIS light, anoxic pool environment) decomposing *P. tremula* leaves in five temporary streams (mean of 4 replicate pool sites) after 10 d (1st sampling, left panel) and approximately 45 d (2nd sampling, right panel). Data were log10(x+1) transformed to adjust scales.

Preconditioning effects of intermittent stream flow on leaf litter decomposition

4.5. Discussion

In this study, we experimentally tested the effects of preconditioning by photodegradation on sunlit streambeds and by anaerobic fermentation in anoxic pools of drying streams on leaf decomposition during the subsequent flowing phase. Overall, results supported our initial hypotheses. We found changes in leaf chemistry, mainly due to anoxic preconditioning, and significant decay retarding effects of preconditioning on European aspen leaves in four of the five temporary streams. Preconditioning effects on decomposer communities, however, were only observed for fungal biomass.

4.5.1. Preconditioning effects on leaf chemistry and leaf mass loss

The effect of preconditioning on leaf chemistry differed between the two types of treatments. Anoxic preconditioning increased the proportion of less palatable refractory organic compounds (lignin, cellulose, and phenolics), which was accompanied by a decrease in the C:N ratio of the leaf tissue. In previous studies, in contrast, the leaching of nutrients led to an increase in the C:N ratio, which in turn reduced the subsequent leaf mass loss rate (Webster and Benfield 1986, Hladyz et al. 2009). In our study, the observed decrease in the C:N ratio was probably due to the intensive leaching and breakdown of labile carbon compounds, which increased the proportion of the remaining refractory carbon compounds (Boulton and Boon 1991). This most likely led to the observed retardation in mass loss, indicating that the quality of carbon may be more important for leaf mass loss than the C:N ratio per se.

Although we detected no effect of irradiation on leaf tissue components, there was a significant decrease in % AFDM. Nonetheless, irradiated leaves tended to lose more nutrients than control leaves during subsequent leaching (chapter 5), suggesting that in addition to leaf desiccation (Leopold et al. 1981)

Preconditioning effects of intermittent stream flow on leaf litter decomposition

photodegradation may also favor the release of nutrients from leaves during inundation, causing a reduced mass loss rate.

The mass loss we observed was comparable to previous reports for congeneric species in permanent streams (e.g. *P. nigra*: Langhans et al. 2008, *P. tremuloides*: Royer and Minshall 2001, see also summary by Casas and Gessner 1999). Unexpectedly, leaves from the first samples in Taibilla and Demnitzer and leaves in fine bags from Vallcebre gained AFDM over time. In Taibilla and Vallcebre, mass loss was probably restricted due to extensive carbonate deposition on leaf surfaces, especially in fine mesh bags, which was consistent with an increase in ash content (i.e. decrease in loss on ignition of leaf material down to 38% LOI in Vallcebre). Although a lower ignition temperature was chosen (450 °C instead of 550 °C) to avoid carbonate combustion, part of the carbonate could have been volatilized and therefore increased the calculated portion of organic compounds, leading to an overestimation in remaining AFDM. However, carbonate coating was not observed in Demnitzer. In this stream, biofilms may have developed on the surface of the leaves, contributing to an increase in AFDM together with organic deposition.

The effect of preconditioning on leaf mass loss also differed between the two types of treatments. Mass loss after approximately 45 days was reduced by 7% AFDM and 4% AFDM when leaves were preconditioned by anoxic incubation and irradiation, respectively. The effect of preconditioning on leaf mass loss was consistent over four of the five selected streams. The exception was Vallcebre, where we did not detect any effects. This inconsistency suggests that the effect of preconditioning may be mediated by characteristics of the stream water other than concentrations of nutrients, for example carbonate. In streams, where significant effects were detected, the retardation of leaf mass loss after preconditioning was most likely due to the observed modification in leaf tissue composition. The effect of anoxic preconditioning was more

Preconditioning effects of intermittent stream flow on leaf litter decomposition

pronounced than was the effect of irradiation. This effect was consistent for both sampling dates, and for fine and coarse mesh bags, indicating that preconditioning can affect leaf decomposition by microorganisms and invertebrates at early and late stages of decomposition.

4.5.2. Preconditioning effects on decomposer communities

The effect of preconditioning on leaf chemistry and leaf mass loss was only partially reflected in the abundance and composition of decomposer communities, with an overall stronger effect on fungi than on invertebrates.

Living fungal biomass was in the lower range of values previously reported (Gessner 2005a, Langhans and Tockner 2006). Anoxic preconditioning significantly reduced fungal biomass in decomposing leaves. The most likely explanation was the inhibition of fungal colonization because of the reduced quality of the carbon substrate. An alternative explanation is that terrestrial fungi, which colonize the leaves when they are attached to the tree, and which are

among the first to grow on submerged leaves (Albrectsen et al. 2010), might have been killed during preconditioning under anoxic, acidic conditions. This may have retarded the growth of fungal biomass until re-colonization. Moreover, the concentration of ergosterol in fungal biomass may differ among fungal species and among their specific physiological states (Salmanowicz and Nylund 1988), which may lead to differences in ergosterol content despite similar biomass. Mass loss in fine mesh bags was significantly reduced by both irradiation and anoxic preconditioning. However, the ergosterol content was only significantly reduced in the anoxic preconditioned leaves. The fungal biomass in irradiated leaves may be similar to control leaves, but a reduced fungal activity may have resulted in slower mass loss.

Preconditioning effects of intermittent stream flow on leaf litter decomposition

Although anoxic preconditioning significantly reduced mass loss in coarse mesh bags, there was no effect on macroinvertebrate abundance. Invertebrates feeding on leaves may indeed have consumed less because of the low chemical quality and reduced fungal colonization of the leaf material, but they may have used the leaf packs not only as a food source but also as a habitat, which might have concealed the feeding effect (Richardson 1992, Dudgeon and Wu 1999). We believe that data on invertebrate biomass could have provided additional information, but assumed that abundance and biomass often follow the same pattern (Dangles et al. 2004).

Differences in mass loss between coarse and fine mesh bags are assumed to result from the different contributions of microbial versus invertebrate communities to total mass loss (Webster and Benfield 1986, Boulton and Boon 1991). We found that invertebrates contributed 32% to total mass loss, with values within previously reported ranges (Graça 2001, Langhans et al. 2008). The highest contribution of invertebrates to mass loss was found in Demnitzer and Vallcebre, which may be due to the higher abundance of shredders, compared to the other streams. In the other streams, collector-gatherers were dominant, while crustacean shredders, in particular, were rare because they are less adapted to intermittent stream flow (Maamri et al. 1997, Langhans and Tockner 2006, Datry et al. 2011). Intermittent stream flow favors organisms with drought resisting strategies, such as diapause and terrestrial stages (Williams 2006), which is supported by the high proportion of insects found in our study.

4.5.3. Implications for organic matter dynamics in temporary streams

Ecosystem processes (such as organic matter decomposition, stream metabolism, and nutrient uptake) occurring during the dry phase in temporary streams have rarely been investigated, and have only recently been

Preconditioning effects of intermittent stream flow on leaf litter decomposition

incorporated into conceptual models of temporary streams (Larned et al. 2010). Larned et al. (2010) proposed that the organic matter is continuously processed during downstream transport in permanent streams. This longitudinal pathway is interrupted by terrestrial processing modes during intermittence of water flow in temporary streams. Organic matter deposited during intermittence of flow is subject to a diversity of processing modes, including decomposition by aquatic and terrestrial communities, anaerobic and aerobic microbial metabolism, leaching, and material degradation by solar radiation and desiccation. Larned et al. (2010) consequently predicted that temporary streams will exhibit (i) longitudinal gradients in processing rates of organic matter with (ii) higher processing rates during flow periods than during dry periods and (iii) an increasing processing efficiency with increasing number of drying and re-flooding cycles due to the high diversity of processing modes.

Our study provides empirical evidence that photodegradation on sunlit stream beds, and the anaerobic processes in residual pools, modify leaf chemistry, which in turn reduces fungal biomass and decelerates leaf decomposition rates after flow recovery. These effects are likely to be enhanced by extended dry events and repeated drying and re-flooding cycles. Supporting the main conceptual ideas of Larned et al. (2010), this indicates that during dry phases, leaf processing is not only reduced but a preconditioning of leaves occurs, which has consequences on their further decomposition during wet phases. Therefore, the temporal and spatial heterogeneity of drying and re-flooding cycles should be taken into account when examining organic matter dynamics in temporary streams. Corroborating this, some previous studies reported decomposition of leaves to be slower at sites with low inundation permanence (e.g. Maamri et al. 1997, Datry et al. 2011), but without any acceleration of the decomposition rate at a higher frequency of drying and re-flooding cycles (Langhans and Tockner 2006). The effect was generally

Preconditioning effects of intermittent stream flow on leaf litter decomposition

attributed to the scarcity of leaf-shredding invertebrates under temporary wet conditions, leading to a decrease in invertebrate-driven leaf processing (Maamri et al. 1997, Langhans and Tockner 2006, Datry et al. 2011, our study). Our study adds the effect of leaf preconditioning during dry phases reducing leaf decomposition rates during wet phases due to leaf chemical modifications. Consequently, the efficiency of organic matter processing in temporary streams is likely low, and extended no-flow periods may lead to an increase in organic matter of poor quality being intermittently transported downstream.

These findings are especially relevant if we consider that the geographical extent of temporary streams is expected to increase worldwide due to the effects of climate and land use change. We expect that in permanent streams that will develop intermittence of flow in the future, the organic matter dynamics will change towards lower decomposition efficiency and discontinuous processing. The accumulation and preconditioning of organic matter during the intermittence of flow will result in major nutrient and carbon pulses at first flush events, and a decrease of substrate quality leading to retardation of organic matter decomposition.

In the present study, we empirically quantified the preconditioning effects on leaf litter decomposition and leafassociated decomposers in temporary streams. Further studies are needed to understand the underlying biogeochemical mechanisms during preconditioning, the effect of preconditioning on microbial community structure and function, the interactions with other in-stream processes such as nutrient uptake and respiration, as well as the consequences for organic matter balances at the stream network and catchment scale.

4.6. Acknowledgments

We are very grateful for the major support by the staff of all chemical laboratories involved, and we particularly thank Angela Krüger of the Institute for Freshwater Ecology and Inland Fisheries in Berlin for the analysis of ergosterol. We are grateful for the comments of T. Datry and three reviewers that helped to improve the manuscript. This study was funded by the EU-FP 7 project MIRAGE (FP7-ENV-2007-1, http://www.mirage-project.eu). We also acknowledge funding from the Spanish Ministry of Science and Innovation (Warmtemp project, CGL2008-05618-C02-01/BOS) for the experiments at the Fuirosos site. D. von Schiller was supported by a fellowship of the German Academic Exchange Service (DAAD) and the *la Caixa* Foundation.

Preconditioning effects of intermittent stream flow on leaf litter decomposition

4.7. Appendix

Table 4.5 List of invertebrate taxa found in leaf litter bags (8 mm mesh size) from a decomposition experiment of *P. tremula* leaves in four temporary streams.

Taxon (family, species)	stream
Ancylidae	Fuirosos
Asellidae	Demnitzer Mühlenfließ
Asellus aquaticus	Demnitzer Mühlenfließ
Baetidae	Candelaro, Fuirosos, Taibilla, Vallcebre
Baetis sp.	Taibilla, Vallcebre
Caenidae	Taibilla
Caenis sp.	Taibilla
Calopterygidae	Candelaro
Calopteryx sp.	Candelaro
Capniidae	Vallcebre
Capnia sp.	Vallcebre
Ceratopogonidae	Demnitzer Mühlenfließ, Fuirosos, Taibilla, Vallcebre
Chironomidae	Candelaro, Demnitzer Mühlenfließ, Fuirosos, Taibilla, Vallcebre
Brillia sp.	Vallcebre
Chironomus sp.	Candelaro
Corynoneura sp.	Vallcebre
Eukiefferiella sp.	Vallcebre
Micropsectra sp.	Vallcebre
Parametriocnemus sp.	Vallcebre
Paratrissocladius sp.	Vallcebre
Rheocricotopus sp.	Vallcebre
Curculionidae	Vallcebre
Diamesinae	Candelaro
Dytiscidae	Candelaro, Demnitzer Mühlenfließ, Fuirosos, Taibilla, Vallcebre
Agabus/Ilybius sp.	Demnitzer Mühlenfließ, Taibilla, Vallcebre
Deronectes sp.	Taibilla
Hydroglyphus sp.	Taibilla
Hygrotus sp.	Vallcebre
Laccophilus sp.	Taibilla, Vallcebre
Elmidae	Fuirosos
Empididae	Candelaro, Taibilla
Ephemerellidae	Fuirosos
Gammaridae	Candelaro, Demnitzer Mühlenfließ
Gammarus sp.	Candelaro, Demnitzer Mühlenfließ
Gammarus pulex	Demnitzer Mühlenfließ
Gammarus roeseli	Demnitzer Mühlenfließ
Glossiphonidae	Demnitzer Mühlenfließ
Glossiphonia sp.	Demnitzer Mühlenfließ
Hydraenidae	Fuirosos
Helophorus sp.	Fuirosos
Hydrophilidae	Taibilla
Laccobius sp.	Taibilla
Heptageniidae	Vallcebre

Preconditioning effects of intermittent stream flow on leaf litter decomposition

Taxon (family, species)	stream
Ecdyonurus sp.	Vallcebre
Hirudinea	Candelaro
Hydracarina	Fuirosos
Leptophlebiidae	Fuirosos, Vallcebre
Leptophlebia sp.	Vallcebre
Lestidae	Fuirosos
Leuctridae	Taibilla, Vallcebre
Leuctra sp.	Taibilla, Vallcebre
Limnephilidae	Demnitzer Mühlenfließ, Fuirosos, Taibilla, Vallcebre
Glyphotaelius pellucidus	Demnitzer Mühlenfließ
Ironoquia dubia	Demnitzer Mühlenfließ
Limnephilus auricula	Demnitzer Mühlenfließ
Limnephilus bipunctatus/ L. centralis	Demnitzer Mühlenfließ
Limnephilus lunatus	Demnitzer Mühlenfließ
Limnephilus rhombicus	Demnitzer Mühlenfließ
Limnephilus stigma	Demnitzer Mühlenfließ
Mesophylax sp.	Taibilla
Stenophylax sp.	Vallcebre
Limnaeidae	Demnitzer Mühlenfließ
Galba truncatula	Demnitzer Mühlenfließ
Nemouridae	Demnitzer Mühlenfließ, Fuirosos, Vallcebre
Nemoura sp.	Demnitzer Mühlenfließ, Vallcebre
Nemoura cinerea	Demnitzer Mühlenfließ
Perlodidae	Fuirosos
Physidae	Taibilla
Physella sp.	Taibilla
Polycentropodidae	Vallcebre
Plectrocnemia sp.	Vallcebre
Psychodidae	Vallcebre
Sericostomatidae	Vallcebre
Simuliidae	Candelaro, Demnitzer Mühlenfließ, Fuirosos, Taibilla, Vallcebre
Simulium sp.	Demnitzer Mühlenfließ
Spionidae	Candelaro
Polydora sp.	Candelaro
Tipulidae	Vallcebre
Vellidae	Fuirosos

5. Light-mediated and anoxic leaf preconditioning at intermittent stream flow affects microbial colonization and mass loss rates

This chapter includes an article, which was reviewed and re-submitted in revised form to Wiley-Blackwell Publishing Ltd.

Dieter, D.; Frindte, K.; Krüger, A. & Wurzbacher, C. (2013) Light-mediated and anoxic leaf preconditioning at intermittent stream flow affects microbial colonization and mass loss rates. Freshwater Biology, re-submitted in revised form.

5.1. Abstract

In intermittent streams, seasonal flow intermittence often coincides with early leaf abscission of riparian vegetation due to water stress. When accumulated on dry stream beds or in remaining pools, leaves are exposed to solar radiation or fermentation processes, respectively. Very little information exists, however, on how these preconditioning processes could affect leaf decomposition when stream flow has recovered. We simulated natural preconditioning of leaves by irradiation with UV-VIS fluorescent lamps and incubation under anoxic conditions. Mass loss rates of preconditioned leaf litter from deciduous trees (*Alnus glutinosa*, *Fraxinus excelsior*, *Populus tremula*, and *Quercus petreae*) were quantified in a temporary stream during base flow conditions. Coarse and fine mesh litter bags were used to study the effect of benthic macroinvertebrates and microorganisms on leaf mass loss. Preconditioning reduced the concentration of macronutrients such as P, K, and Mg and increased the relative cellulose content of the leaves. Preconditioning changed the fungal community structure (analyzed by DGGE) depending on leaf species and sampling date. Preconditioning in anoxic conditions also

Light-mediated and anoxic leaf preconditioning at intermittent stream flow affects microbial colonization and mass loss rates

suppressed living fungal decomposer biomass (measured as ergosterol) by 42% resulting in 33% lower mass loss rates in fine mesh bags. In contrast, mass loss rates were not affected by preconditioning when macroinvertebrate decomposers had access to the leaf litter. In streams exhibiting seasonal flow intermittence, preconditioning will influence organic carbon dynamics towards lower rates of microbially-mediated turnover and towards poorer quality of downstream-transported material.

5.2. Introduction

Temporary streams are characterized by seasonal intermittence of stream flow. Water abatement periods typically show contraction of the channel or complete desiccation, while standing ponds can persist in pool sections until water abatement is complete (Steward et al. 2012). Temporary streams are a global phenomenon (Larned et al. 2010) and their spatial extent as well as the duration of ceased flow are expected to increase due to alterations in climate, land-use, and water withdrawal (Tockner et al. 2009). Streams that are currently perennial may become temporary in the future. Although temporary streams are a global phenomenon with recent topicality, these aquatic systems are not well understood. Uncertainties remain in the prediction and evaluation of ecological processes in temporary and incipient temporary streams.

One of the most important processes in forested streams is organic material turnover, which encompasses various aspects of aquatic ecology, such as physicochemical water properties, metabolism, and decomposer communities (Gessner et al. 1999). Most notably in headwater streams, terrestrial leaf litter of a riparian forest represents a major carbon source for heterotrophic stream communities (Vannote et al. 1980). Yet, it is poorly understood how conditions during flow intermittence affect the decomposition kinetics of leaves and the composition and dynamics of the associated decomposer community (aquatic

Light-mediated and anoxic leaf preconditioning at intermittent stream flow affects microbial colonization and mass loss rates

hyphomycetes, macroinvertebrates). In response to water stress, seasonal leaf abscission may occur earlier (Acuña et al. 2007) and leaves accumulate on dry sediment surfaces or in remaining ponds until being decomposed when stream flow recovers. Slower mass loss rates of leaves were observed on dry sediment surfaces when compared to those in remaining ponds, where mass loss was still slower than in the flowing channel (Maamri et al. 1997). Leaves that accumulate on dry sediment surfaces are exposed to intense desiccation and solar radiation. This might photochemically degrade leaf compounds (Day et al. 2007) that can be rapidly lost upon rewetting, before decomposer communities could utilize them. The results reported by Dieter et al. (2011) indicated a reduced decomposability of irradiated aspen leaves after inundation. In remaining ponds, accumulation of organic material leads to high concentrations of dissolved organic carbon (DOC) and nutrients, while pH levels decline and oxygen is depleted (Lake 2003, Canhoto and Laranjeira 2007, Schlief and Mutz 2011). Leaves that are trapped in these ponds are exposed to intense leaching of nutrients and labile carbon and fermentation processes, which degrade the leaf chemical composition (Küsel and Drake 1996, Reith et al. 2002).

The chemical composition of leaf tissue is known to influence its decomposition rate and usually varies among leaf species (Webster and Benfield 1986). Higher nutrient concentration supports decomposition of leaf material, whereas higher levels of recalcitrant or enzyme inhibiting C compounds suppress decomposition (Zhang et al. 2008). Hence, the modification of the chemical composition of leaves by preconditioning during the dry phase could in turn affect leaf decomposing communities and thus leaf mass loss rates when stream flow recovers. These processes have not been fully elucidated and information about microbial colonization of leaf litter in intermittent streams in general is rare (Artigas et al. 2011).

Light-mediated and anoxic leaf preconditioning at intermittent stream flow affects microbial colonization and mass loss rates

The aim of our study was to investigate the indirect effects of seasonal flow intermittence in streams on the decomposition of leaf litter of different quality. We presumed that abscission of riparian leaf litter during the dry season leads to a preconditioning of the leaf material by photodegradation on dry river beds and fermentation processes in anoxic ponds. We hypothesized that this preconditioning reduces the quality and therefore palatability of the leaf substrate. Colonization by microbial and macroinvertebrate decomposers was therefore hypothesized to be inhibited and their community structure to be altered, which would reduce the leaf mass loss rate after stream flow has recovered.

5.3. Methods and Material

5.3.1. Preconditioning of leaves

In autumn 2009, we installed nets to collect fallen senescent leaves from the deciduous tree species *Alnus glutinosa* (L.) Gaertn., *Fraxinus excelsior* L., *Populus tremula* L., and *Quercus petraea* (Mattuschka) Liebl. These species cover a range of different chemical composition (Table 5.1) and a range of common species of riparian vegetation. Freshly collected leaves were air-dried by ventilation with a fan in the laboratory. Leaf preconditioning (Figure 5.1) under anoxic conditions was performed by incubating 100 g of leaves of each species in 40 L of artificial lake water (Fischer et al. 2006) under anoxic (dissolved oxygen < 0.2 mg L^{-1}) and acidic (pH ~ 5.0) conditions at room temperature (~ 20 °C). Aquatic microorganisms were introduced by inoculation with stream water (250 mL). Leaves were removed after 21 d of incubation and quickly air-dried by ventilation with a fan in the laboratory. Photodegradation of a second set of leaves was implemented by irradiation with UV and daylight fluorescent lamps (Cosmedico Arimed B6 [with 31% UVB of total UV], Osram Biolux 965; 50 W m^{-2} total radiation, 17 W m^{-2} UV radiation; measured with

Light-mediated and anoxic leaf preconditioning at intermittent stream flow affects microbial colonization and mass loss rates

LiCor 1800 spectroradiometer) for 12 h daily over a period of 21 d. A control set of leaves was stored under dark and dry conditions.

Table 5.1 Tissue components (mean ± 1SD, n = 4) of air-dried senescent leaves of four deciduous tree species. AFDM: ash free dry mass. Ratios are based on percentages.

	Alnus glutinosa	*Fraxinus excelsior*	*Populus tremula*[a]	*Quercus petreae*
AFDM (%)	93.4 ± 0.2	90.1 ± 0.4	90.8 ± 0.0	94.0 ± 0.3
phenolics (%)	6.1 ± 0.6	2.8 ± 0.4	3.2 ± 0.3	5.6 ± 0.3
cellulose (%)	20.6 ± 1.3	26.3 ± 1.3	24.1 ± 0.6	29.6 ± 0.8
lignin (%)	18.6 ± 3.2	13.5 ± 1.5	14.5 ± 0.6	22.1 ± 2.1
C (%)	47.3 ± 0.4	43.3 ± 0.3	46.5 ± 0.1	46.1 ± 0.3
N (%)	2.0 ± 0.0	1.5 ± 0.0	0.8 ± 0.0	1.0 ± 0.0
C:N ratio	23.3 ± 0.7	29.1 ± 0.4	61.3 ± 2.6	48.4 ± 1.3
lignin:N ratio	9.2 ± 1.7	9.1 ± 0.9	19.0 ± 0.6	23.1 ± 1.9
P (mg g^{-1} DM)	0.7 ± 0.1	0.8 ± 0.1	0.8 ± 0.1	1.9 ± 0.1
Ca (mg g^{-1} DM)	16.6 ± 0.8	26.6 ± 1.3	25.4 ± 0.2	9.4 ± 1.0
Mg (mg g^{-1})	1.4 ± 0.1	4.9 ± 0.3	1.6 ± 0.1	1.4 ± 0.1
K (mg g^{-1} DM)	7.1 ± 1.2	1.9 ± 0.5	3.2 ± 0.4	4.8 ± 0.9

[a] Selected data from *P. tremula* were originally published in Dieter et al. 2011

The chemical composition of preconditioned leaves was determined for each treatment group and the control in replicates of four subsamples. In order to estimate nutrient and mass loss by initial inundation, a 24 h leaching cycle was performed with four subsamples of control and irradiated leaves. Leaf tissue components were then compared with samples from the anoxic preconditioning, where leaves had already leached during incubation. Leachates were filtered through 0.7 μm glass fiber filters (Whatman GF/F) and analyzed for dissolved nutrients (see below).

Subsamples of freshly collected leaves and leached preconditioned leaves were lyophilized, ground to fine powder in a ball mill, and analyzed for contents of total carbon and nitrogen (Elementar vario EL C/N elemental analyzer, Hanau, Germany), fibrous compounds (gravimetric determination of acid-detergent lignin and cellulose, Gessner 2005a), and phenols (detection with a spectrophotometer following extraction with 70% acetone and addition of Folin-Cialteau reagent, Bärlocher and Graça 2005).

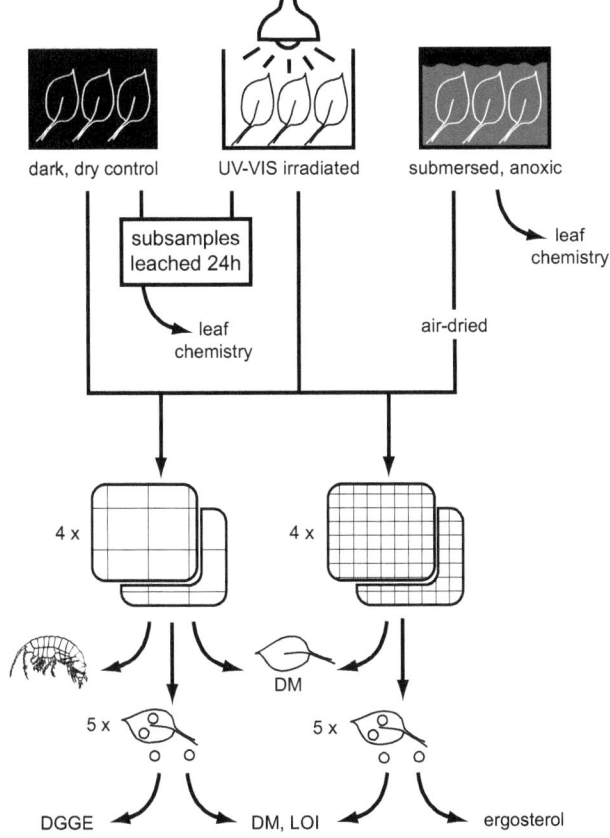

Figure 5.1 Laboratory procedure for preconditioning of leaves for 21 d (storage at dark and dry conditions, irradiation with UV-VIS light, incubation in anoxic water) and in-stream immersion using coarse and fine mesh bags that were sampled at two dates (after 10 d and again after 20, 31, 38, or 40 d depending on the leaf species). DM: dry mass, LOI: loss on ignition, DGGE: denaturing gradient gel electrophoresis.

Furthermore, concentrations of Ca, Mg, K, and P were determined by ignition (4 h at 450 °C) followed by extraction with 10 mL of aqua regia:ultra-pure water

Light-mediated and anoxic leaf preconditioning at intermittent stream flow affects microbial colonization and mass loss rates

(1:1) solution in an ultra-sonic bath (Väisänen et al. 2008) and detection with an ICP-OE spectrometer (Thermo Scientific, iCAP 6500, USA, wavelengths: Ca 318.1 nm axial, K 766.4 nm axial; Mg 285.2 nm axial, P 214.9 nm radial).

Chemical composition changes in preconditioned leaves compared with control leaves were assessed using student's t-test comparing each treatment with the control.

5.3.2. Field experiments

Leaf mass loss experiments were performed in the lowland temporary stream Demnitzer Mühlenfließ (52° 21' N, 14° 11' E), a 3rd order tributary to the Spree River, NE Germany (Gelbrecht et al. 2005). In the 38 km^2 catchment of the Demnitzer Mühlenfließ, mean annual precipitation is 480 mm and mean air temperature is 9 °C. The experimental site runs through a dense deciduous riparian forest, but upstream agricultural sites supply high nutrient inputs, especially during high flow. During the experiments, water temperature was low, but increased with time of the experiment, and nutrient concentrations were moderate (Table 5.2).

For the mass loss experiments, packs of ~ 4 g of preconditioned leaves were placed into coarse mesh (8 mm) and fine mesh (0.5 mm) nylon bags (Bärlocher 2005). The experiments were carried out in March/April 2010. At four replicate pool sites, leaf litter bags were tied to an iron rod, and anchored in the bed of the stream. Bags were first sampled after 10 d and again after 31 d (*A. glutinosa*), 20 d (*F. excelsior*), 38 d (*P. tremula*), and 40 d (*Q. petraea*), respectively, depending on advancement of decomposition. Bags were cut off, placed in individual plastic bags and transported to the laboratory on ice in a cooler.

During the experiment, water temperature was monitored in hourly intervals with an iButton temperature logger (iBCod 22L, Alpha Mach Inc., Mont St-

Light-mediated and anoxic leaf preconditioning at intermittent stream flow affects microbial colonization and mass loss rates

Hilaire, Canada) fixed to an iron rod at each of the four replicate sites. Dissolved oxygen (mg L^{-1}), pH, specific conductance ($\mu S\ cm^{-1}$), and current velocity (cm s^{-1}) were measured with hand-held sensors at each site on the day of immersion of leaf bags and at the leaf bag removal dates after 10, 20, 31, and 38 days (n = 20). In addition, surface water was collected at each site, filtered (0.7 µm glass fiber filters, Whatman GF/F) and stored in polyethylene bottles for analysis of dissolved nutrients.

Table 5.2 Values (mean ± 1SD, n = 20) of the main physical and chemical water characteristics during a leaf mass loss experiment in the temporary stream Demnitzer Mühlenfließ. DO: dissolved oxygen, DIC: dissolved inorganic carbon, DOC: dissolved organic carbon, DN: total dissolved nitrogen, SRP: soluble reactive phosphorus, DP: total dissolved phosphorus.

Water temperature (°C)[a]	3.9 ± 3.2
Current velocity (cm s^{-1})	20 ± 9
Depth (cm)	33 ± 10
specific conductance ($\mu S\ cm^{-1}$)	903 ± 32
pH	8.2 ± 0.1
DO (mg L^{-1})	11.2 ± 1.3
Cl^- (mg L^{-1})	33.4 ± 3.5
SO_4^{2-} (mg L^{-1})	142.4 ± 15.5
DIC (mg C L^{-1})	66.3 ± 5.0
DOC (mg C L^{-1})	13.5 ± 0.8
NO_3^- (mg N L^{-1})	9.4 ± 1.7
NH_4^+ (mg N L^{-1})	0.13 ± 0.03
DN (mg N L^{-1})	10.4 ± 2.2
SRP (µg P L^{-1})	47 ± 3
DP (µg P l^{-1})	55 ± 4

[a] mean of data recorded at hourly intervals

Water samples from the field and from leaf leachates were immediately analyzed for total dissolved phosphorus (DP) and soluble reactive phosphorus (SRP) concentrations with a spectrophotometer (VARIAN Photometer CARY 1E)

Light-mediated and anoxic leaf preconditioning at intermittent stream flow affects microbial colonization and mass loss rates

and for dissolved inorganic carbon (DIC) with a C/N elemental analyzer (Jena Analytik, multi N/C 3100, Jena, Germany). Samples were stored at -20 °C until further analysis of dissolved organic carbon (DOC) and total dissolved nitrogen (DN, both measured with C/N elemental analyzer). In addition, ammonium was measured with a spectrophotometer (Skalar SCAN++, Netherlands), and concentrations of nitrate, chloride and sulfate were quantified using ion chromatography (Shimadzu, Japan).

Leaf samples from field experiments were processed immediately upon arrival at the laboratory (Figure 5.1). Individual leaves were carefully washed with tap water and macroinvertebrates from coarse mesh bags were preserved in 70% ethanol until individuals were counted, identified, and assigned to functional feeding groups (Schmidt-Kloiber et al. 2006). Invertebrate abundances were expressed as number of individuals per g DM of leaves. Tests were performed on log10(x+1) transformed data to approach normal distribution. Differences in abundances of taxa or feeding groups between preconditioned and control samples and between samples of different leaf species were assessed by ANOSIM using PRIMER software (version 6).

A set of two discs was cut (11 mm diameter with a cork borer) from each of five randomly selected leaves from every bag (central vein avoided). The first set of five discs was placed in an aluminum pan and oven-dried at 40 °C (48 h) together with the leaf pack to obtain oven-dry mass (DM). Discs were subsequently ignited at 450 °C for 4 h to obtain loss on ignition (LOI). Total remaining ash free dry mass (AFDM) of every leaf pack was then calculated by multiplying LOI with the sum of DM of twice the five discs and the leaf pack. Initial 100% AFDM remaining was calculated using LOI from subsamples of preconditioned, leached leaves and was corrected for handling loss during transport and deployment by use of additional control samples carried to the field and back at the day of deployment (Benfield 2006). Linear leaf mass loss rate k' (dd^{-1}) was calculated as %AFDM lost divided by cumulative degree days of in-stream immersion (Petersen and Cummins 1974). Overall effects of leaf species (n = 4), mesh size (n = 2), sampling date (n = 2), and preconditioning (n = 3) on mass loss rates were tested using Kruskal-Wallis test.

The second set of five discs was stored at -20 °C prior to analysis of (i) the ergosterol content to assess living fungal biomass (from fine mesh bags) and (ii) the microbial community structure (from coarse mesh bags, see below). Total ergosterol content was measured through microwave-assisted liquid phase extraction (Young 1995) and detection with HPLC and UV spectrometry (Gessner 2005b, Dionex UltiMate 3000 LC, Sunnyvale CA, USA). The ergosterol content was expressed as µg ergosterol per g DM of leaves.

Differences in mass loss rate, macroinvertebrate abundance, and ergosterol content between preconditioned and control samples were assessed using Wilcoxon signed rank or rank sum tests, as appropriate. Relations among variables were estimated using Pearson's product moment correlation with data from the second sampling when macroinvertebrate abundance and ergosterol content peaked and leaf decomposition was at an advanced stage. These statistical analyses were performed using R statistical software with the significance level set to α = 0.05 for all tests.

5.3.3. Microbial community structure

DNA extraction, PCR amplification, and denaturing gradient gel electrophoresis (DGGE) techniques were used to analyze the bacterial and fungal community structure in leaf discs. To compare leaf species and treatment effects on microbial communities, all leaf-species were compared after 10 d of incubation in the stream. Additionally, one leaf species (*P. tremula*) was chosen for monitoring microbial succession during a time series of the decomposition process (before incubation, day 10, day 38).

Light-mediated and anoxic leaf preconditioning at intermittent stream flow affects microbial colonization and mass loss rates

The DNA of three leaf discs per sample was extracted by using chloroform-phenol-isoamylalcohol and zirconium beads following the protocol of Nercessian et al. (2005). The pellet was dissolved in 40 μL DNase-free water and stored at -20°C. Two different primer sets were used in PCR amplifications for 16S rDNA and functional gene groups (Table 5.3). PCR mix contained 0.4 pmol of each primer, 250 μM of each desoxyribonucleoside triphosphate, 3 mM MgCl, 5 μL 10× PCR buffer, 2 μL of bovine serum albumin (20 mg mL^{-1}), 1 μL of template DNA and 0.5 U BIOTAQ Red DNA polymerase (Bioline) in a total volume of 50 μL.

We used DGGE to detect changes in microbial community composition in preconditioned leaves (Muyzer et al. 1993). The PhorU system was used for each DGGE amplification in this experiment (Ingeny, Goes, Netherlands). The polyacrylamide content and the gradient of urea and formamide varied between the different primer systems (Table 5.3). The gels were stained with SybrGold. For sequencing, the DGGE bands were excised and dissolved in 20 μL TE buffer. Re-amplification products were retrieved by using 2 μL template DNA. PCR products were purified by 1:1 addition of a precipitation solution (20% polyethylenglycol 8000 and 2.5 mol L^{-1} NaCl in destilled water). PCR products were incubated for 20 min and then washed with 100 μl 70% ethanol. All sequencing was performed by using a sequencer (Applied Biosystems, Darmstadt, Germany) following manufacturers sequencing protocol.

The ARB software (http://www.arb-home.de) was used for phylogenetic analyses of 16S rRNA bacteria sequences. The retrieved 16S sequences were pre-aligned by using the SINA alignment service (http://www.arb-silva.de/aligner) imported in a comprehensive 16S ARB database and subsequently corrected manually. Sequences were then added to the tree according to maximum parsimony criteria. The BLAST analysis of fungal DNA sequences identified similarity scores between isolates.

Table 5.3 Primer systems and DGGE conditions as used for the analysis of bacterial and fungal community structure on decomposing leaf litter in a temporary stream.

Primer	Function	Sequence	Program	Reference	DGGE gradient polyacryl-amide
341f 803r 907r	16S Bacteria	CCT ACG GGA GGC AGC AG CCG TCA ATT CMT TTG AGT TT CCG TCA ATT CMT TTG AGT T	10 min at 95°C; 1 min at 95°C, 1 min at 55°C, 1 min at 72°C (35 cycles); 10 min at 72°C	Muyzer et al. 1993, Lee et al. 1993, Teske et al. 1996	40-65 7%
189f[a] 654r	16S ammonium oxidizers	GGA GRA AAGYAG GGG ATC G CTA GCY TTG TAGTTT CAA ACG C	5 min at 95°C; 45 s at 95°C, 45 s at 58°C, 45 s at 72°C (40 cycles); 10 min at 72°C	Kowal-chuk et al. 1997	40-65 7%
nirK laCuf nirK3Cur*	Nitrite reductase (Cu depending, nirK)	ATC ATG GTS CTG CCG CG GCC TCG ATC AGR TTG TGG TT	10 min at 95°C; 45 s at 95°C, 45 s at 65°C, 45 s at 72°C (40 cycles); 10 min at 72°C	Throback et al. 2004	40-70 8%
Arc344f Arc915r	16S archaea	TCG CGC CTG CTG CIC CCC GT GTG CTC CCC CGC CAA TTC	5 min at 94 °C; 1 min at 94 °C (19 cycles), 1 min at 71 °C (−0.5 °C per cycle), 2 min at 72 °C; 1 min at 94 °C (20 cycles), 1 min at 61 °C , 2 min at 72 °C ; 10 min at 72 °C	Raskin et al. 1994 Casama-yor et al. 2002	
ITS3f* ITS4r	ITS2 fungi	GCA TCG ATG AAG AAC GCA GC TCC TCC GCT TAT TGA TAT GC	10 min at 95°C; 30 s at 95°C, 30 s at 55°C, 1 min at 72°C (35 cycles); 10 min at 72°C	Raviraja et al. 2005	20-80 8%

[a] GC clamp: CGC CCG CCG CGC CCC GCG CCC GTC CCG CCG CCC CCG CCC G

Differences in microbial community structure were tested by analysing DGGE gels in GelCompar II (Applied Maths, Keistraat, Belgium). Resulting banding patterns were exported as binary matrix and subsequent statistics were performed using R statistical software (version 2.12.0, Ihaka and Gentleman 1996, www.r-project.org) and vegan package for multivariate statistics (vegan package, V 1.17-10, Oksanen et al. 2011). Communities were plotted as NMDS plots using the function metaMDS and Kulczyński as distance index. Leaf species parameters were fitted into the plots by using a significance level of 0.001 (Attend the non-linear nature of NMDS.) and class centroids were calculated as ellipses based on a confidence level of 0.95. Correlations between matrices were calculated by a Mantel test with 5000 permutations based on Spearman rank correlations. Separation of leaf species, preconditioning treatments, and time series were furthermore supported by a permutated ANOVA (betadiv, adonis) and analysis of variance homogeneity (betadisper) with 1000 permutations based on betadiversity Arrhenius index matrices (Oksanen et al. 2011).

5.4. Results

5.4.1. Preconditioning of leaves

Preconditioning significantly affected the chemical composition of leaves. During irradiation, leaves lost about 2% of their initial DM (data not shown). When previously irradiated, all but *A. glutinosa* leaves leached more nutrients and DOC (Wilcoxon signed rank, $W = 59.5$, $P < 0.05$, $n = 24$) (Table 5.4). After preconditioning including leaching, the proportion of leaf constituents differed significantly between preconditioning treatments and control (Figure 5.2). However, differences were inconsistent among species and treatments, but most pronounced for *F. excelsior* leaves and for the anoxic preconditioned leaves. Effects of irradiation only (data not shown) were mostly concealed by the following leaching. Therefore, only *A. glutinosa* and *F. excelsior* were significantly affected showing a decrease in Ca, Mg, K, P, N, and lignin content. The proportion of N, cellulose, and lignin tended to increase during anoxic preconditioning and a decrease of the C:N and lignin:N ratios was observed. Anoxic preconditioning over 21 d also resulted in high leaching losses for total phenols (up to 50%), K (~ 90%), P (up to 70%) and Mg (up to 80 %) compared to freshly collected, air-dried leaves (Table 5.1).

Table 5.4 Amount of nutrients leached from packs of air-dried leaves (mg element leached per g dry weight of leaves) during 24 h of incubation in artificial stream water at 20°C in mesocosms. Previously to leaching, control leaves were stored under dark and dry conditions and irradiated leaves were exposed to UV and daylight fluorescent lamps. SRP: soluble reactive phosphorus, DP: dissolved phosphorus, DN: dissolved nitrogren, DOC: dissolved organic carbon.

	Alnus glutinosa		*Fraxinus excelsior*		*Populus tremula*		*Quercus petreae*	
	control	irradiated	control	irradiated	control	irradiated	control	irradiated
	mg g^{-1}							
SRP - P	0.20	0.20	0.10	0.14	0.21	0.25	1.01	1.16
DP - P	0.21	0.21	0.10	0.21	0.22	0.26	1.03	1.17
DN - N	1.18	0.88	0.83	1.07	0.73	1.31	0.24	0.34
NH_4^+ - N	0.17	0.04	0.03	<0.01	0.04	0.02	<0.01	0.10
NO_3^- - N	0.01	0.08	0.07	0.03	0.06	0.16	<0.01	0.09
DOC - C	54.86	48.53	63.77	66.67	73.85	75.77	12.41	17.99

5.4.2. Leaf associated microbial communities

Ergosterol concentrations of field samples were low after 10 d in the stream averaging 24 µg g^{-1} DM (Figure 5.3). At the second sampling date ergosterol concentration ranged between 58 and 497 µg g^{-1} DM with highest values for *Q. petreae* leaves, which remained longest immersed in the stream until the last day of the experiment (40 d). Leaves from the anoxic preconditioning had 42% and 31% lower ergosterol concentrations than the control and irradiated leaves, respectively (Wilcoxon signed rank, anoxic - control: $W = 103$, $P < 0.001$, $n = 16$; anoxic - irradiated: $W = 13$, $P < 0.05$, $n = 16$). No significant difference was detected between irradiated and control leaves. Increases in ergosterol content were positively correlated with higher mass loss ($r = 0.80$, $P<0.001$), but did not correlate with leaf chemical compounds.

Light-mediated and anoxic leaf preconditioning at intermittent stream flow affects microbial colonization and mass loss rates

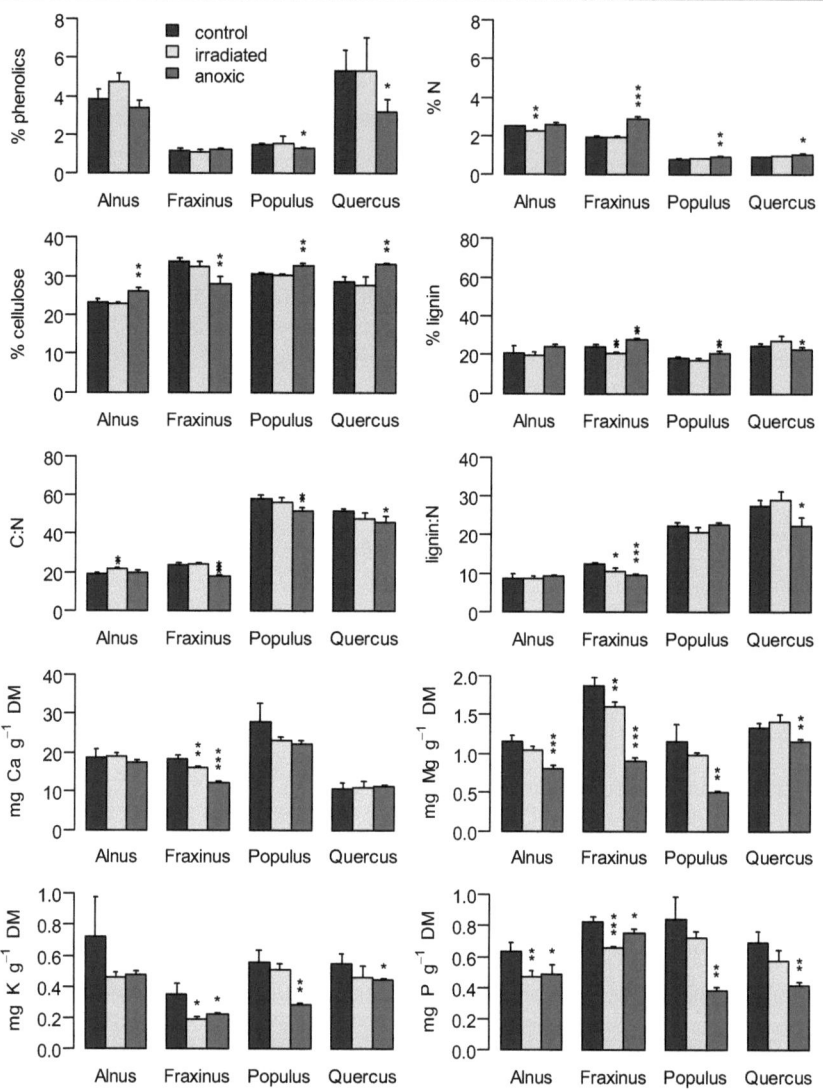

Figure 5.2 Proportions of chemical compounds in the leaf tissue of four species (mean ± SD, n = 4). Leaves were leached for 24 h following storage at dark and dry conditions (control) and irradiation with UV-VIS light for 21 d or incubated in anoxic water for 21 d. Asterisks indicate significant differences between treatment and the control (Student's t-test, *** $P < 0.001$, ** $P < 0.01$, * $P < 0.05$). DM: dry mass of leaves.

Light-mediated and anoxic leaf preconditioning at intermittent stream flow affects microbial colonization and mass loss rates

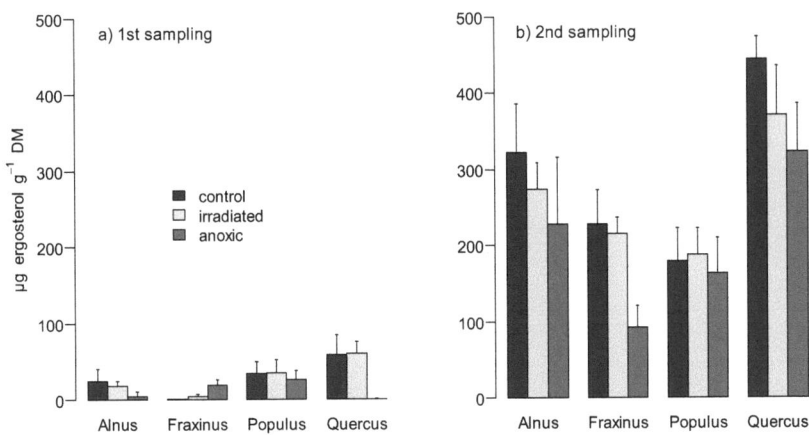

Figure 5.3 Concentration of ergosterol (mean ± SD, n = 4) in leaves of four species immersed in a stream for a) 10 d (1st sampling) and b) 20 to 40 d (2nd sampling) depending on the leaf species. Prior to immersion, leaves were preconditioned by irradiation with UV-VIS light and by incubation under anoxic conditions or stored dark and dry (control). DM: dry mass of leaves.

Observed differences in fungal community structure were remarkable among leaf species separated by certain leaf tissue components (Figure 5.4a). A global preconditioning effect was only visible for leaves from the anoxic treatment associated with the second axis (Figure 5.4b). Differences between irradiated and control leaves were inconsistent among leaf species and had no effect after 10 d of incubation in the stream (see also statistics in Table 5.6 in chapter 5.7 Appendix). However, the time series on *P. tremula* leaves showed a weak effect of the irradiation treatment on the fungal community during the incubation period in the stream, whereas the initial effect of the anoxic preconditioning tended to diminish towards the second sampling after 38 d (Table 5.6 in chapter 5.7 Appendix, Figure 5.7 in chapter 5.7 Appendix). The fungal community consisted of typical species found on leaves decomposing in streams (aquatic hyphomycetes, Figure 5.6 in chapter 5.7 Appendix).

Light-mediated and anoxic leaf preconditioning at intermittent stream flow affects microbial colonization and mass loss rates

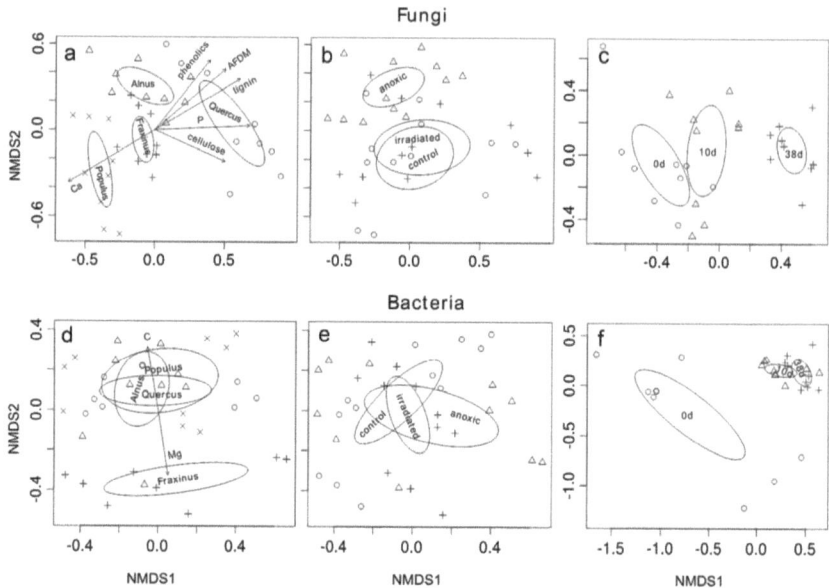

Figure 5.4 NMDS plots based on fungal (upper panel) and bacterial (lower panel) DGGE community profile. Ellipses calculated on a 95% confidence level are shown around their centroids for a) & d) leaf species, b) & e) preconditioning treatments, and c) & f) sampling dates for the time series of *P. tremula* leaves. Leaf tissue components that were potentially underlying the separation of leaf species were fitted into the plots a) & d). Stress values are given for each plot.

The bacterial community on *F. excelsior* leaves differed significantly from the other leaf species, seemingly explained by C and Mg contents (Figure 5.4d, second NMDS axis). The effect of leaf preconditioning on the bacterial community structure was negligible and only observed in certain leaf species (Table 5.6 in chapter 5.7 Appendix). Time series of *P. tremula* revealed that treatment effects occurred only before immersion, being replaced by a markedly different community in the stream (Figure 5.4f, Table 5.6 in chapter 5.7 Appendix). Bacterial communities were dominated by typical freshwater *Beta-* and *Alphaproteobacteria* (Figure 5.6 in chapter 5.7 Appendix).

Light-mediated and anoxic leaf preconditioning at intermittent stream flow affects microbial colonization and mass loss rates

Ammonium oxidizers on leaves retrieved from the stream were absent or very scarce (data not shown), but a surprisingly high diversity of denitrifying bacteria was found (34 bands in total). Remarkably, DNA from neither nitrite reductase K (nirK) nor *Archaea* was detected in the preconditioned leaf samples before immersion in stream water (independent of preconditioning treatment), but were detected on all the samples after immersion in the stream (data not shown). This contrasted with bacterial 16S (and fungal ITS) genes, which could be readily detected in all examined samples.

Potential pairwise correlations across microbial community patterns (bacteria, fungi, nirK) were not significant for the leaf species comparison (10 d), but revealed a high correlation of nirK community with bacterial and fungal community patterns in the time series of *P. tremula* leaves (10 d, 38 d) (pairwise Mantel test, nirK - fungi: $r = 0.501$, $P < 0.001$; nirK - bacteria: $r = 0.362$, $P < 0.001$).

5.4.3. Leaf associated macroinvertebrate communities

Average abundance of invertebrates in leaf packs ranged between 17 and 368 Ind g^{-1} DM (Figure 5.5) and was about 59 Ind g^{-1} DM higher on the second sampling day than on the first sampling day (Wilcoxon signed rank, $W = 0$, $P < 0.001$, n = 48). On the first sampling day, macroinvertebrates appeared in limited numbers in only 24 of 48 leaf bags. Abundances differed among leaf species (ANOSIM, global R = 0.243, $P = 0.001$), but were generally not affected by leaf preconditioning.

Light-mediated and anoxic leaf preconditioning at intermittent stream flow affects microbial colonization and mass loss rates

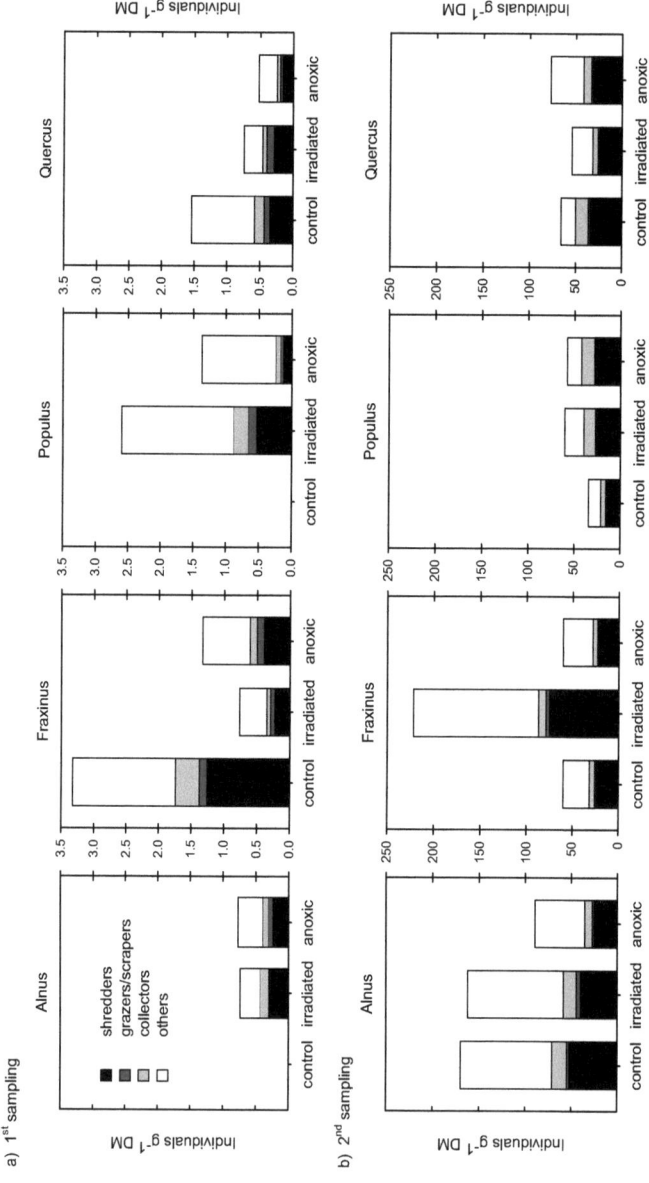

Figure 5.5 Abundance of macroinvertebrates assigned to selected feeding groups colonizing leaves of four species in coarse mesh litter bags deployed in a stream for a) 10 d and b) 20 to 40 d (depending on leaf species). Prior to immersion, leaves were preconditioned by irradiation with UV-VIS light and by incubation under anoxic conditions or stored dark and dry (control). DM: dry mass of leaves.

Light-mediated and anoxic leaf preconditioning at intermittent stream flow affects microbial colonization and mass loss rates

5.4.4. In-stream leaf mass loss

Leaf mass loss was relatively fast, but differed between sampling dates (Kruskal-Wallis, χ^2 = 134.55, P < 0.001, n = 191). During the first 10 d in the stream, an increase in AFDM (up to 9%) was observed for all leaf species, which was probably due to deposition of organic debris (although leaves were carefully washed) or due to colonization by microbial decomposers. For that reason, mass loss rates were only calculated for the second sampling date (Table 5.5).

Table 5.5 Linear leaf mass loss rate k' (mean ± 1SD, n = 4) for 4 leaf species in coarse and fine mesh bags as calculated from the second sampling (< 50% leaf mass remaining in coarse mesh bags) in the stream Demnitzer Mühlenfließ. Prior to immersion, leaves were preconditioned by irradiation with UV-VIS light and by incubation under anoxic conditions or stored dark and dry (control).

		k' (%AFDM dd^{-1})	
		coarse mesh bag	fine mesh bag
Alnus glutinosa	control	0.41 ± 0.04	0.10 ± <0.01
	irradiated	0.45 ± 0.01	0.12 ± 0.01
	anoxic	0.45 ± <0.01	0.09 ± <0.01
Fraxinus excelsior	control	0.43 ± 0.17	0.16 ± 0.03
	irradiated	0.56 ± 0.21	0.10 ± 0.03
	anoxic	0.45 ± 0.20	0.08 ± 0.03
Populus tremula	control	0.21 ± 0.04	0.11 ± <0.01
	irradiated	0.24 ± 0.01[a]	0.06 ± 0.02
	anoxic	0.21 ± 0.04	0.09 ± 0.02
Quercus petreae	control	0.26 ± 0.02	0.07 ± 0.01
	irradiated	0.21 ± 0.11	0.07 ± 0.02
	anoxic	0.27 ± 0.04	0.05 ± <0.01

[a] n=3

Overall leaf mass loss rates differed among species (Kruskal-Wallis, χ^2 = 10.9, P < 0.05, n = 95) and were 74% lower in fine mesh compared to coarse mesh bags (Kruskal-Wallis, χ^2 = 64.4, P < 0.001, n = 95). No overall effect of treatment was detected, but interactions occurred. Significant treatment effects were only detected for fine mesh bags (Wilcoxon signed rank, irradiation: W = 110, P < 0.05, n = 16; anoxic preconditioning: W = 133, P < 0.001, n = 16). The

Light-mediated and anoxic leaf preconditioning at intermittent stream flow affects microbial colonization and mass loss rates

irradiation treatment reduced mass loss rates by 43% on average for *F. excelsior* and *P. tremula*, and increased mass loss rates by 18% for *A. glutinosa*. The anoxic preconditioning reduced mass loss rates by 28% on average over all leaf species. Differences in mass loss between the two preconditioning treatments were insignificant.

5.5. Discussion

Photodegradation and anoxic incubation enhanced the loss of nutrients during leaching, while the proportion of poorly degradable compounds increased in the leaf tissue. In-stream effects were particularly strong following the anoxic incubation, with reduced biomass of living fungi, altered microbial community structure, and inhibited microbially mediated leaf mass loss. However, mass loss rates were not affected when macroinvertebrate decomposers participated in leaf breakdown.

5.5.1. Preconditioning of leaves

High losses of DOC, P, N, K, and Ca from leaf tissue during leaching detected in our study were comparable to the results from Gessner (1991). Proportions lost were species-specific, thus partially balancing differences in initial leaf chemical composition. *Q. petraea*, for example, had a high initial P concentration, but after leaching its P concentration was within the range of the other leaf species. Leaf leaching seems to have a homogenizing effect on leaf quality. This indicated that the proportion of nutrients remaining after first leaching rather than initial chemical composition of leaves should be considered when predicting mass loss rates from leaf chemical composition.

Desiccation of leaves, as it occurred through air-drying in our study, can enhance leaching loss compared to leaching of fresh leaves (Gessner 1991, Taylor and Bärlocher 1996). In temporary streams, desiccation of leaves is very likely to occur during the dry phase (Acuña et al. 2007) and the effect of desiccation can be intensified by photodegradation. Degrading effects of UV and VIS radiation were reported to induce carbon dioxide release and higher dissolved organic matter release from leaves mainly due to breakdown of lignin and lipids (Anesio et al. 1999, Day et al. 2007, Brandt et al. 2009). We demonstrated that irradiation not only led to a direct mass loss, but enhanced leaching of DOC, supporting previous studies (Denward and Tranvik 1998, Anesio et al. 1999) and also led to enhanced nutrient losses, suggesting that a pulse of high DOC and nutrient release upon re-flooding is driven by additive effects from both desiccation and photodegradation. Very recently, it was shown that terrestrial aging of leaf litter enhanced leaching loss of nutrients and DOC (Fellman et al. 2013). Although the effects faded with duration of litter aging (probably due to rain-mediated interim leaching), higher leaching loss was still detectable after 4 to 6 months of aging. Leaching losses increased significantly in anoxic incubated leaves when compared to previously irradiated leached leaves, suggesting that leaching is not restricted to the first 24 h of submersion. We observed a steady increase of P concentration in the water tanks (Figure 6.3), reaching saturation only after approximately 8 d, while DOC concentrations peaked after 2 d and then remained nearly constant. This suggests that leached labile DOC components are available only shortly after re-flooding, whereas nutrients are available for longer time spans. In standing ponds however, prolonged leaching of nutrients was therefore shown to degrade leaf litter quality with poorly decomposable compounds remaining.

5.5.2. Leaf associated microbial communities

In many bags of the second sampling there was not enough leaf material left to perform all microbial analyses. Fungal biomass was therefore only analyzed for the fine mesh bag samples, while the microbial community structure was analyzed for coarse mesh bag samples. However, to test for treatment effects, it was not important to analyze leaf samples from both mesh sizes, although microbial communities may have differed among bags of different mesh sizes.

Degradation of leaf material by preconditioning reduced living fungal biomass and microbially-mediated mass loss rates. Fungal biomass was estimated by measuring ergosterol concentration. However, ergosterol content of fungi is known to vary depending on the fungal species and its physiological state (Gessner and Chauvet 1993). Hence, low ergosterol concentrations may have been the result of (i) actual inhibition of colonization and growth of fungi on the leaves or (ii) an effect of leaf colonization by different fungal species that are less selective or more specialized in the preconditioned substrate, but contain less ergosterol. The latter would be supported by the results from the DGGE analyses showing different fungal community structures for preconditioned leaves, especially for the anoxic preconditioning. Some fungi may have died or were inhibited under acidic and anoxic conditions. In addition, this may have altered the leaves to conditions that favor a second group of fungal decomposers, the aero-aquatic hyphomycetes, which are adapted to low oxygen conditions and could therefore gain importance in temporary streams (Bärlocher et al. 1978). In contrast to bacteria, fungal community structure still differed among preconditioned leaves even at later stages of in-stream decomposition, pointing to a certain level of maintenance of the initial community (Sridhar et al. 2009). However, similar mass loss rates were observed for differing fungal communities (Sridhar et al. 2009, Ferreira and Chauvet 2012) implying that a change in fungal community does not necessarily change functionality. Therefore, it seems likely that growth of the fungal community was indeed inhibited by the anoxic preconditioning.

Besides preconditioning, inhibition of fungi may have been additionally caused by bacteria. Although bacteria in decomposing leaf litter usually

Light-mediated and anoxic leaf preconditioning at intermittent stream flow affects microbial colonization and mass loss rates

account for less than 10% of microbial biomass (Baldy et al. 1995), bacteria are known to have a pronounced negative effect on fungal communities and can therefore suppress fungal biomass and mass loss rates (Romaní et al. 2006). In fact, actinomycetes (*Actinoplanes*, *Kineospora*) were detected in this study, which are typical antagonists to fungi (Das et al. 2012).

The effect of UV-VIS irradiation on microbial community structure by affecting fungi that were already in the leaves during preconditioning was presumably marginal, due to protection from irradiation by the leaf matrix (for fungi growing inside the leaf tissue) or due to incorporation of melanin and carotenoids in fungal cell walls (Nosanchuk and Casadevall 2003, Libkind et al. 2003). Denward et al. (2001) showed that fungal biomass was reduced during irradiation of submersed leaf litter, but was unaffected during irradiation of dry leaf litter. Hence, fungi seem to be less affected directly by the irradiation, but may be inhibited by the DOM leached from the litter as being observed after initial drying and irradiation of leaves in our study.

The analysis of the bacterial community structure provided not only the detection of relevant taxa but also a first screening of a functional gene (nirK) and functional phylogenetic groups (ammonium oxidizers, *Archaea*) on decomposing leaf litter. Leaf chemistry had only a weak structuring effect on bacterial community composition, which is in accordance with earlier findings (Das et al. 2012). Our data suggests that leaves of four different leaf species are colonized by the same stream-water bacteria upon immersion in stream water leading to a complete replacement of the initial communities. Moreover, this colonization seems to be independent of (i) leaf-litter species, with the exception of *F. excelsior*, (ii) preconditioning of the leaves, and (iii) duration of submergence within a range of 38 days (Figure 5.4f).

The high diversity of nitrite reductase genes appearing in all leaf samples is remarkable, indicating that species using this alternative (anaerobic) respiratory pathway (denitrification step reducing nitrite to NO) are omnipresent regardless of anoxic preconditioning. Our data indicated a coupling between the fungal and the denitrifying community, which may reflect direct interaction or dissolved nitrogen as shared resource. No evidence for such a direct positive interaction between fungi and bacteria regarding nitrogen was found or reported in previous studies so far.

5.5.3. Leaf associated macroinvertebrate communities

Neither total abundance nor taxonomic composition of the invertebrate community was affected by the preconditioning of leaves. Further, mass loss rates in coarse mesh bags were similar among preconditioning treatments. This suggests that overall mass loss rates of leaves in temporary streams are not affected by preconditioning when macroinvertebrate decomposers are abundant. However, feeding preferences or discrimination between leaves was not observed directly. Individuals either may have used the leaf packs only as habitat or may have actually not discriminated between different qualities of the food source. However, non-selective feeding and the consumption of leaf litter with low nutritional quality, e.g. due to reduced fungal biomass, can reduce growth and growth efficiency of macroinvertebrates (Albariño et al. 2008).

5.5.4. In-stream leaf mass loss

Our study showed that leaf species with high lignin:N ratios (*P. tremula*, *Q. petraea*) were decomposed at slower rates than those with low lignin:N ratios (*A. glutinosa*, *F. excelsior*) confirming previous findings (Loranger et al. 2002). However, although lignin:N ratios declined, mass loss rates were lower for the anoxic preconditioned leaves compared to the control leaves. This may have been the result of the observed increase in the proportion of recalcitrant

C (lignin, cellulose) due to the loss of labile carbon and nutrients during leaching. Reduced mass loss rates of leaves from the anoxic preconditioning suggested that the observed changes in leaf chemistry provided substrates of poor quality and poor palatability for microbial decomposer communities. Irradiation of leaves had a weaker and only partial influence on the mass loss rate, likely because microorganisms were less affected by the modified leaf chemistry. Dieter et al. (2011) detected similar effects in a range of temporary streams. While the authors reported the effects to be independent of the stream water characteristics, our results documented the effects of preconditioning by UV-VIS irradiation to be leaf species-dependent. For example, no enhanced leaching of irradiated *A. glutinosa* leaves was observed and irradiation effects led to an increase in mass loss, which contrasted effects on leaching and mass loss for *F. excelsior* and *P. tremula* leaves. *A. glutinosa* and *Q. petreae* differed from the other two species by a higher initial content of phenolics and lignin. Lignins were reported to be decomposed by photodegradation (Austin and Ballaré 2010), which facilitates microbial access to other compounds that are occluded by lignin. Hence, the degradation of the recalcitrant lignin compounds was likely to have leveled the inhibiting effects of nutrient loss resulting in unaffected (*Q. petreae*) or even accelerated (*A. glutinosa*) mass loss rates.

5.5.5. Conclusion

Our data suggest that preconditioning of leaves during the dry period of temporary streams is leaf species-dependent, but generally results in poor quality and reduced microbially-mediated mass loss rate of leaf litter. It therefore affects one of the most important carbon substrates for headwater stream communities. Future flow intermittency may influence organic carbon dynamics towards slower rates of microbial turnover of allochthonous C and towards poorer quality of downstream-transported material. Fungal decomposer communities will be effectively altered by the changed substrate quality, but will be finally determined by the conditions of the re-flooding water.

5.6. Acknowledgements

We thank Sarah Schmarsow, Thomas Rossoll, Elke Zwirnmann, Martin Winter, Daniel Graeber, and Sandra Hille for their help with the field work and the laboratory analyses. We thank Kirsten Pohlmann for statistical advice. The comments of Thomas Mehner (*scientific writing* course at the IGB), Klement Tockner, and two anonymous reviewers very much improved the manuscript. We thank Ivan J. Grimmett for professional language corrections. This study was funded by the EU-FP 7 project MIRAGE (FP7 ENV 2007 1, www.mirage-project.eu). D. Dieter was supported by a fellowship for specialized courses of the German Academic Exchange Service (DAAD, D/10/40525).

Light-mediated and anoxic leaf preconditioning at intermittent stream flow affects microbial colonization and mass loss rates

5.7. Appendix

Table 5.6 Statistics on microbial community structure in different decomposing leaf species at different sampling dates and from different preconditioning treatments. na: not applicable, *F*: *Fraxinus excelsior*, *Q*: *Quercus petraea*.

Test	Organism	Dataset	Samples tested	Method	dF	Residuals	F-model	r2	Pr	Homogeneity of groups	Tukey HSD	Tukey p adjusted
Treatment effects within leaf-species												
treatment	bacteria	leaf spec.	Quercus	default	2	8	29.572	0.9	0	0.966	na	na
treatment	fungi	leaf spec.	Quercus	default	2	8	2.852	0.5	0.02	0.08	na	na
treatment	nirk	leaf spec.	Quercus	default	2	7	1.145	0.3	0.4	0.805	na	na
treatment	bacteria	leaf spec.	Alnus	default	2	8	3.446	0.5	0.02	0.726	na	na
treatment	fungi	leaf spec.	Alnus	default	2	8	2.938	0.5	0.02	0.43	na	na
treatment	nirk	leaf spec.	Alnus	default	2	8	1.744	0.4	0.07	0.585	na	na
treatment	bacteria	leaf spec.	Fraxinus	default	2	8	10.052	0.8	0	0.117	na	na
treatment	fungi	leaf spec.	Fraxinus	default	2	8	4.322	0.6	0.01	0.265	na	na
treatment	nirk	leaf spec.	Fraxinus	default	2	8	0.661	0.2	0.86	0.172	na	na
treatment	bacteria	leaf spec.	Populus	default	2	8	17.512	0.9	0	0.39	na	na
treatment	fungi	leaf spec.	Populus	default	2	8	3.768	0.6	0.01	0.512	na	na
treatment	nirk	leaf spec.	Populus	default	2	7	3.228	0.6	0.01	0.321	na	na
Global treatment effects												
treatment	bacteria	leaf spec.	all	nested	2	35	2.187	0.1	0.01	0.063	na	na
treatment	fungi	leaf spec.	all	nested	2	35	0.129	0.1	0	0.084	na	na
treatment	nirk	leaf spec.	all	nested	2	33	0.843	0.1	0.52	0.661	na	na
Single treatments bacteria/fungi												
treatment	bacteria	leaf spec.	c – UV	nested	1	23	1.158	0.1	0.02	0.333	na	na
treatment	fungi	leaf spec.	c – UV	nested	1	23		0	0.13	0.711	na	na
treatment	bacteria	leaf spec.	c – anox	nested	1	23	2.775	0	0.02	0.186	na	na
treatment	fungi	leaf spec.	c – anox	nested	1	23	3.582	0.1	0	0.101	na	na

Light-mediated and anoxic leaf preconditioning at intermittent stream flow affects microbial colonization and mass loss rates

Test	Organism	Dataset	Samples tested	Method	dF	Residuals	F-model	r2	Pr	Homogeneity of groups	Tukey HSD	Tukey p adjusted
Global seperation by leaf species												
leaf spec.	bacteria	leaf spec.	all	nested	3	35	4.883	0.3	0	0.544	na	na
leaf spec.	fungi	leaf spec.	all	nested	3	35	0.312	0.3	0	0.018	Q,F	0.014
leaf spec.	nirK	leaf spec.	all	nested	3	33	2.38	0.2	0	0.674	na	na
Coupled effects of leaf species and treatment												
leaf spec.: treatment	bacteria	leaf spec.	all	default	6	35	10.474	0.4	0	na	na	na
leaf spec.: treatment	fungi	leaf spec.	all	default	6	35	2.862	0.2	0	na	na	na
leaf spec.: treatment	nirk	leaf spec.	all	default	6	33	1.559	0.2	0.02	na	na	na
Time series on P. tremula leaves												
Global treatment effects												
treatment	bacteria	time series	all	nested	2	17	1.345	0.2	0.1	0.356	na	na
treatment	fungi	time series	all	nested	2	17	2.102	0.2	0	0.792	na	na
treatment	nirK	time series	all	nested	2	17	2.031	0.2	0	0.478	na	na
Single treatments fungi/nirK												
treatment	fungi	time series	contr – UV	nested	1	11	2.274	0.2	0.01	0.607	na	na
treatment	nirK	time series	contr – UV	nested	1	11	2.126	0.2	0.01	0.377	na	na
treatment	fungi	time series	anox – contr	nested	1	11	2.154	0.2	0.03	0.972	na	na
treatment	nirK	time series	anox – contr	nested	1	11	4.166	0.3	0.01	0.213	na	na
Separation of time points (days 10 & 38)												
time	bacteria	time series	all	nested	1	17	5.028	0.2	0.01	0.807	na	na
time	fungi	time series	all	nested	1	17	5.837	0.3	0	0	Okt 38	0.0005
time	nirK	time series	all	nested	1	17	27.441	0.6	0	0.41	na	na

Light-mediated and anoxic leaf preconditioning at intermittent stream flow affects microbial colonization and mass loss rates

Test	Organism	Dataset	Samples tested	Method	dF	Residuals	F-model	r2	Pr	Homogeneity of groups	Tukey HSD	Tukey p adjusted
Coupled effects of leaf species and treatment												
time: treatment	bacteria	time series	all	default	2	17	12.941	0.4	0	na	na	na
time: treatment	fungi	time series	all	default	2	17	7.257	0.3	0	na	na	na
time: treatment	nirK	time series	all	default	2	17	7.918	0.1	0	na	na	na

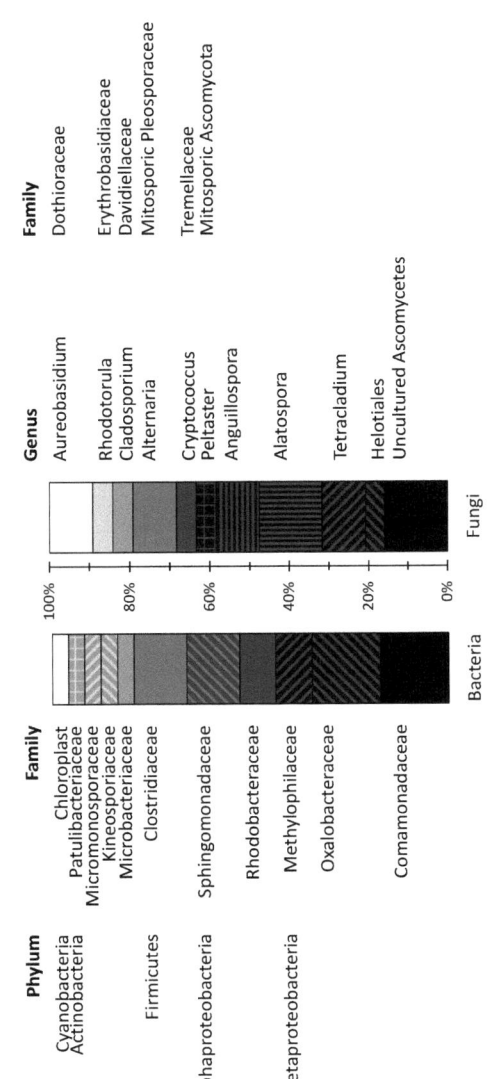

Figure 5.6 Bacterial and fungal communities on decomposing leaf litter in the intermittent stream Demnitzer Mühlenfließ, NE Germany.

Light-mediated and anoxic leaf preconditioning at intermittent stream flow affects microbial colonization and mass loss rates

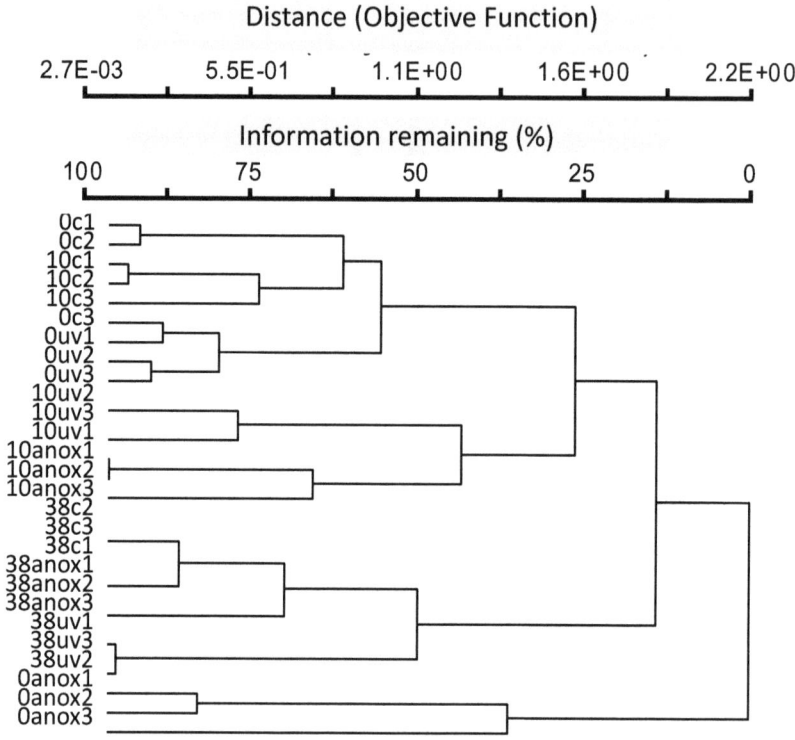

Figure 5.7 Cluster analysis for fungal communities on preconditioned (c : control, uv: UV-VIS irradiated, anox: incubated in anoxic water; replicates 1,2,3) *P. tremula* leaves after 0 d, 10 d, and 38 d of decomposition in the intermittent stream Demnitzer Mühlenfließ, NE Germany.

6. Phosphorus dynamics in sediments – An overview

6.1. The aquatic phosphorus cycle and phosphorus forms

Phosphorus (P) is commonly known as the dominant limiting nutrient for primary production and microbial activity in freshwater ecosystems. P occurs in inorganic and organic forms, with each form being either in a dissolved or particulate state. The natural inorganic P input into freshwater ecosystems originates from precipitation, particulate atmospheric deposition, and from weathering of rocks containing P, such as apatite ($Ca_5(PO_4)_3OH$, Schwoerbel 1999). The concentration of phosphate (main dissolved inorganic P form) relevant for direct biological uptake is typically measured colorimetrically after reaction with molybdate (Murphy and Riley 1962). It is therefore commonly referred to as soluble reactive P (SRP). Non-reactive P represents basically organic P compounds and polyphosphates, although some very labile organic P forms may be detected as SRP. Natural concentrations of SRP in water are usually very low (<< 0.1 mg L^{-1}), but can be largely increased by anthropogenic inputs from point sources (e.g. sewage discharge) and non-point sources (e.g. fertilizers) (Mainstone and Parr 2002). As a result of anthropogenic inputs, eutrophication of freshwater systems has been widely observed (Schwoerbel 1999). P inputs to lakes usually have relatively long residence times in the system, while P is rapidly transported in fluvial systems. P transported in rivers represents the largest flux of the global P cycle (Liu et al. 2008). The annual amount is estimated at 17.7 - 30.4 x 10^{12} g total P, 55% of which are the result of human activities (Compton et al. 2000).

Due to relatively long residence times, the mobilizable P pool in the sediment of lakes can be more than three orders of magnitude larger than in the overlying water body. The concentrations of SRP in the sediment pore water usually account for only 0.5 - 5% of the total P in the sediment (Hupfer 1995).

There is no clear relationship between total P content of the sediment and the trophic state of an aquatic system, but the proportion of bioavailable and mobilizable P forms were often reported to be more relevant (Hupfer 1995). In addition, functional characteristics such as morphology, water residence time, catchment lithology, microbial activity, external inputs, and dominant physicochemical conditions play an important role in determining the trophic state of an aquatic system. Sediments are in general a P sink and can retain large amounts of external P inputs. However, they can turn into net P sources in eutrophied systems when the pressure of external inputs of P is drastically reduced. P accumulated in the sediment is then gradually released, which has often resulted in the maintenance of the eutrophic state for several years (hysteresis effect).

Cycling of P in aquatic systems is characterized by physical, chemical, and biological mechanisms (Boström et al. 1988). These involve processes in the water column, the sediment, the pore water, and in biota. The principal processes include (i) sorption and desorption to and from particles, (ii) precipitation and dissolution, (iii) assimilation, transformation, and release, (iv) mineralization of organic matter, (v) sedimentation and re-suspension, (vii) as well as advective and diffusive transport (Figure 6.1).

Phosphorus dynamics in sediments – An overview

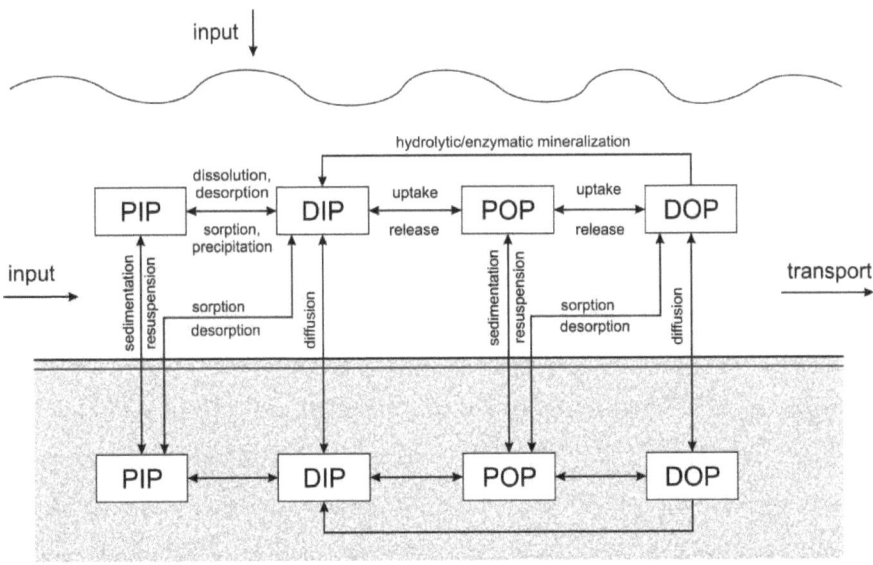

Figure 6.1 Conceptual model of phosphorus (P) cycling in the sediment (grey-shaded) and overlying water column of aquatic systems. PIP: particular inorganic P, DIP: dissolved inorganic P, POP: particular organic P, DOP: dissolved organic P.

P can be associated with or sorbed to suspended or sedimentary particles such as humic acid-metal-complexes, clay, Ca, Mn, as well as Fe and Al oxyhydroxides until being desorbed again. In addition, phosphate can be fixed in Ca, Mn, Fe, and Al minerals until dissolution. There is a gradual transition between adsorption and precipitation, and thus both processes are usually summarized to the terms *abiotic P uptake* or *abiotic P release*, respectively. Oxidized forms of Fe and Mn are known to have a much higher binding affinity for P than their reduced forms. Classical theories of Einsele (1936, 1938) and Mortimer (1941, 1942) therefore related P release mainly to the reduction of ferric iron under anaerobic conditions. However, this paradigm is known to represent a special case of P mobilization, because P can also be released from sediment surfaces if there are aerobic conditions in the hypolimnion (Hupfer and Lewandowski 2008).

Dissolved forms of labile organic and inorganic P can be taken up and assimilated by autotrophic as well as heterotrophic biota and move as organic P through the food chain until release or mineralization (Allan and Castillo 2007). Under aerobic conditions and high P concentrations, some bacterial groups are capable of incorporating even more P than necessary for optimum growth (luxury uptake, Reddy et al. 1999). This P is stored intracellular by formation of polyphosphates to ensure P supply at low P concentrations in the water column. Up to 31 - 50% of the non-reactive P in surface sediment layers was found to be in the form of polyphosphates incorporated in microorganisms. Polyphosphates are released under anaerobic conditions - a mechanism, which is widely utilized as biological P elimination during waste water treatment (Hupfer and Gächter 1995, Baldwin and Mitchell 2000, Hupfer et al. 2007). Microbial P uptake can account for up to 91% of total P retention in a woodland stream or up to 60% in wetlands (Reddy et al. 1999). High assimilation rates can cause low P concentrations in the water, which support desorption of P from particles into the water. P in living biomass together with detrital P forms represent the pool of organic P, which accounts for 10 to 70% of total P in sediments (Reddy et al. 1999). Easily decomposable organic P forms, such as sugar phosphates, nucleic acids, phosphonates, phospholipids, and ATP, can be distinguished from refractory organic P, such as phytates, and P in humic and fulvic substances (Reddy et al. 1999, Worsfold et al. 2008). Intracellular P may return to the water via excretion, cell lysis, and mineralization of dead organic material.

Particulate P from autochthonous and allochthonous production, atmospheric deposition, and catchment erosion settles to the bottom of water bodies. However, lacustrine primary production is largely recycled within the water column, with only small proportions (20% in Bloesch and Uelinger 1990) reaching the surface of the sediment. Further, about 70% of the total P that reaches the sediment surface can be quickly returned to the water column (Hupfer et al. 1995). Settled material can therefore be relatively depleted in P concentration, but may still release P to the pore water when buried.

Concentration gradients between pore water and the overlying water column induce diffusion of dissolved P along the gradient to approach concentration equilibrium. A steeper concentration gradient leads to higher diffusion rates. In contrast, an increase in sediment bulk density decreases the pore volume and therefore diffusive transport owing to the increase in collisions of dissolved P with solid particles in the sediment and longer pathways (tortuosity). Much faster than diffusive transport, P can be transported with advective water movement due to turbulences. Advection can be caused by seepage due to groundwater discharge, internal waves induced by wind, gas ebullition, and bioturbation caused by organisms (Boström et al. 1988). Burrowing macroinvertebrates inhabiting sediments can promote interstitial and surface water exchange by pumping and sediment relocation. This could markedly affect exchange between oxic and anoxic layers of the sediment. For example, chironomid larvae pumping water through the burrowed tubes promote transport and release of pore water P to surface water or on the contrary promote P uptake by transport of oxygenated surface water into the sediment. Burrow walls become oxidized resulting in a higher P binding affinity (Lewandowski and Hupfer 2005). In addition, macroinvertebrate assemblages in the sediment also contribute to P release by organic matter mineralization and by feeding and defecating activities.

Phosphorus forms in sediments are commonly characterized by sequential P fractionation techniques, which functionally separate P forms by their extractability in specific chemical reagent solutions (Pettersson et al. 1988).

Among these are e.g. neutral, reducing, alkaline, and acidic reagents separating (i) labile bound P from (ii) redox-sensitive Mn and Fe bound P, (iii) hydroxide exchangeable Fe and Al bound P, (iv) organic P, (v) acid-sensitive Ca and carbonate-bound P, and (vi) recalcitrant, occluded P. The nature of the different reagents reflects the fact that redox conditions and pH levels are very important environmental factors controlling P dynamics (chapter 6.2). Boström et al. (1988) proposed main characteristics of lakes as functionally deducible from the dominant extractable P fractions. Systems with increased sewage load, saturated P binding places, and an enhanced potential for P release were characterized by proportionally high concentrations of labile and reductant-soluble P compounds. Sediments with high P uptake affinity due to high Fe and especially Al contents were characterized by a high proportion of alkaline-extractable P forms. Systems without additional P inputs showed a high proportion of acid-extractable P, which is typically lithogenic, apatitic P. However, in more calcareous systems P can be bound to Ca in secondary, non-apatitic forms that can also be easily dissolved in NH_4Cl and acidic extracts (Hupfer 1995). In oligotrophic systems with a very low total P and inorganic P load, labile and biologically available P is rapidly recycled, leaving a small proportion of refractory organic P and major proportions of residual P. Conclusively, the dominance of certain P forms and related processes determine whether the sediment could either act as an important sink or source of P. This information can therefore facilitate the estimation of P release or retention potentials and thus provide important background for management decisions (Hupfer 1995).

6.2. Uptake and mobilization mechanisms of phosphorus in sediments

The dynamics of P uptake and release in sediments involve various biotic and abiotic processes that are part of P cycling (chapter 6.1). The affinity and capacity of sediments for P uptake and sedimentary P mobilization depend on various parameters. The main driving factors are mineralogy, grain size, organic matter content, temperature, pH, redox conditions, and concentrations of oxygen, nitrate, and sulfate (Reddy et al. 1999, Dittrich et al. 2013). In addition, P concentrations in the pore water play a major role for uptake and release. At low P concentrations, sediments tend to release P and vice versa. The concentration at which uptake and release are at equilibrium can be deduced from experimental sorption isotherms that give information about the affinity and capacity of sediments for P uptake (Nair et al. 1984). These methods involve shaking of sediment in a range of P solutions and are therefore also particularly useful to assess sorption dynamics of re-suspended particles. However, they preclude the effects of diffusion and sediment stratification, thus experiments with intact sediment columns provide an additional estimate of P uptake in bottom sediments (Reddy et al. 1999).

The pH value has contrasting effects on P dynamics depending on the main P controlling minerals. In Fe-rich and Al-rich watersheds, an elevated pH level may reduce the P binding capacity of Fe and Al hydroxides due to ligand exchange with hydroxide ions. This can be induced by temporary high primary production in sunlit shallow lakes and stagnant ponds (Boström et al. 1988). In marked contrast to that, increased pH immobilizes P in calcareous systems due to adsorption of P to calcite or binding in Ca phosphate, which can be favored by increased temperatures (Boström et al. 1988). An increase in the pH level due to high primary production can lead to a biogenic calcite precipitation with co-precipitation of P. In contrast to this, an increase in temperature in aphotic zones also enhances respiration and thus the production of CO_2. This, in turn, lowers the pH level and therefore favors carbonate and Ca phosphate dissolution and hence P mobilization. Enhanced metabolic rates caused by higher temperatures can also foster additional P mobilization by mineralization of organic P, which is then partly incorporated into biomass and partly released as inorganic phosphate to the pore water. Dissolved P can be easily released to the overlying water column due to enhanced diffusion rates at higher temperatures. Moreover, high respiration rates result in oxygen depletion and even anoxic conditions. This may in turn affect other uptake/release processes, for example the sorption of P to Fe and Mn minerals, which is strongly redox-sensitive. Under oxic conditions, P is stably bound to amorphous Fe(III) oxyhydroxides. In contrast, at low redox potentials (Eh < 200 mV) ferric iron (Fe^{3+}) will be reduced to ferrous iron (Fe^{2+}), which has a lower P binding ability (Roden and Edmonds 1997). Mn is even more redox-sensitive. At low oxygen concentrations, Mn^{4+} can already be reduced to Mn^{2+} and release associated P (Ostendorp and Frevert 1979).

As a consequence, enhanced oxygen consumption by respiration as occurring for example in stagnant ponds leads to acidification and lowering of the redox potential, which could mobilize both Fe-bound P and carbonate/Ca-associated P. Moreover, microbial production is restricted under anaerobic conditions so that less P can be taken up in microbial biomass (Boström et al. 1988). Some bacterial groups will even release P from hydrolyzed polyphosphates (chapter 6.1). Consequently, Fe-bound P (reductant-soluble P) and polyphosphates (alkaline-extractable non-reactive P in a single step extraction) both represent very redox-sensitive fractions for P mobilization (Hupfer 1995, Hupfer et al. 2007).

However, sediments may not be highly reduced, although they are already anoxic (Baldwin et al. 2000). When oxygen is depleted, specialized bacteria also utilize nitrate, ferric iron or sulfate as electron acceptors, which gradually decreases the redox potential. On the one hand, increased nitrate concentrations can therefore stabilize P binding by preventing anaerobic P release or by inhibiting ferric iron reducing enzymes (Boström and Petterssen

1982). Yet, on the other hand, nitrate can also stimulate mineralization and thus P release (Boström et al. 1988, Jensen and Andersen 1992). The initiation of sulfate respiration produces sulfide, which itself is capable of reducing ferric iron. Moreover, it can substitute P from ferrous iron compounds, due to the much lower solubility of iron sulfide (Sperber 1958, Baldwin et al. 2000, Zak et al. 2006). In fact, a linear correlation between sulfate reduction and P mobilization was reported by Roden and Edmonds (1997). The effect can even be accelerated by higher temperatures, which stimulate sulfate respiration (Boström et al. 1988, Zak et al. 2006). Respiration of sulfate usually occurs at a redox potential (60 – 100 mV), which is lower than that of iron reduction. Yet, iron sulfide partly persists when the redox potential rises again (Boström et al. 1988) leading to a hysteresis controlled mobilization of P by the decoupling of the Fe-P cycle. A large proportion of sedimentary P release can be explained by loss of P from the reductant-soluble P fraction due to the reduction and fixation of Fe and the decoupling of the Fe-P cycle (Hupfer 1995). The concentration of sulfate in aquatic systems may be increased due to anthropogenic sources, which would weaken Fe-related P uptake capacity and accelerate P mobilization (Loeb et al. 2008).

Aeration of sediments facilitates P uptake in the sediment by oxidation of Fe compounds. Although oxygen concentrations in fine sediments drop quickly within the first millimeters or even micrometers below the surface, even low oxygen concentrations can be sufficient to maintain an oxidized layer that serves as a diffusion barrier at the sediment-water interface. It prevents upward diffusion of dissolved P from the pore water of deeper layers by uptake in Fe compounds. Therefore, total P content is usually higher in the surface layers decreasing with depth and pore water P concentrations usually follow a pattern opposite to that of total P with lowest concentrations at the sediment-water interface due to exchanges with the sediment and overlying water column (Reddy et al. 1999). Less oxygen is necessary for the maintenance of an oxidized layer than for the oxidation of a reduced layer (Schwoerbel 1999) turning this layer into an effective barrier against sedimentary P release to the overlying water column. However, the barrier layer will be covered by settling material and turns anoxic when buried, which leads to re-mobilization of the accumulated P. The newly formed barrier layer on top is exposed to increased input of P from pore water as well as from the overlying water column. This may soon lead to a saturation of the P binding capacity depending on the thickness of the barrier layer and the Fe content. Saturation causes the loss of the barrier function and ultimately results in P release. Aeration of sediments therefore does not facilitate long-term P retention, but leads only to a translocation of P release (Gächter 1987).

The amount of Fe present in sediments dictates the amount of sedimentary P uptake and retention. Some studies specified a critical Fe:P ratio, above which further P release from sediments is restricted (Fe:P > 15 for lakes in Jensen et al. 1992, Fe:P > 10 for peatlands in Zak et al. 2010).

6.3. Phosphorus dynamics and water level fluctuations

The mentioned biotic and abiotic processes of P cycling retard the longitudinal P transport in riverine systems and affect its vertical movement in lacustrine systems. As a result, nutrient cycling and transport can be described as a spiral (Newbold et al. 1983). One spiral is defined as the distance travelled by a nutrient while performing one complete cycle from the moment of release, to uptake, and until being released again. The distance travelled in solution until uptake (uptake length) as well as the time and distance a nutrient remains immobilized (turnover time/distance) are commonly used to quantify nutrient spiraling and retention properties. Uptake lengths are usually shorter for a lower hydraulic radius like in low order streams, but are also dependent on geology, sediment grain size, trophic state, and other physical, chemical, and biological factors (chapter 6.2). The effect of climate change, however, is not well understood (Withers and Jarvie, 2008). On the one hand, more severe droughts may cause lower discharge and increase the retardation of downstream P transport due to the decrease in uptake lengths and increase in local storage. On the other hand, flood events following droughts may result in longer uptake lengths and a net P loss. The net effects on annual P fluxes are therefore difficult to assess. P retention was reported to be altered by water level fluctuations of a Mediterranean intermittent stream (von Schiller et al. 2008). Uptake lengths of phosphate were found to increase with discharge, while uptake rates increased with decreasing discharge upon drying. My own preliminary samplings corroborate this concept by showing that P released by a first flush in an advancing water front was more efficiently removed at low-speed advance of the water front (Figure 6.2).

Although nutrient release due to extreme drawdown was also observed in studies on water reservoirs (Fabre 1988, Baldwin et al. 2008), the effects of water level fluctuations or complete drawdown on processes in the sediments are not fully understood for neither riverine nor lacustrine systems. While drying stimulates respiration and mineralization of organic material due to aeration, desiccation is known to kill microorganism, which release nutrients by cell lysis (Baldwin and Mitchell 2000). In addition, rapid rewetting may cause cell disruption by osmotic shock (Turner and Haygarth 2001). As a consequence, significant amounts of P can be released upon re-flooding of sediments. Repeated drying and re-flooding selects for bacteria capable of surviving in drought resistant stages or capable of facultative anaerobic metabolism (Baldwin and Mitchell 2000). Sulfate reducers, for instance, are obligate anaerobic, but can persist in drought resisting stages (Baldwin et al. 2000).

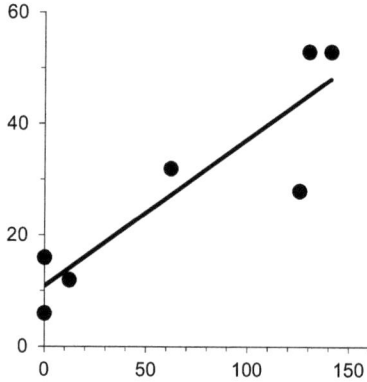

Figure 6.2 Relationship between the speed of an advancing water front and the increase in soluble reactive phosphorus (SRP) concentration of the stream water between a fixed point and the water front during surface flow recovery in a temperate intermittent stream (November 2009, Demnitzer Mühlenfließ, eastern Germany).

Desiccation was also observed to shift the composition of P fractions in sediments and marshland soils (de Groot and Fabre 1993). An increase in Fe(III) oxyhydroxide-bound P due to oxidation of the surface layer of the drying sediment was observed, whereas changes in Ca-bound P were inconsistent but tended to decrease (Moore and Reddy 1994, Baldwin 1996, Kerr et al. 2010). Moreover, an enhanced mineralization of acid-extractable organic P was observed, which was therefore assumed to be the most important organic P fraction.

Studies performed on soils and wetland sediments revealed that repeated drying of sediments had an effect on mineral morphology. Amorphous Fe(III) oxyhydroxides were found to transform into more crystal structures due to repeated desiccation (Lijklema 1980, Qiu and McComb 2002). This process is called *aging* of the minerals and leads to a reduction of sorption places. As a result, P sorption affinity declines and P may be released upon re-flooding (Lijklema 1980, Baldwin et al. 2000). The aging of P binding partners during drying therefore contributes to a first flush of dissolved P, which can cause high P concentrations during the first flood event (Baldwin and Mitchell 2000, Qiu et al. 2004, Wilson and Baldwin 2008, Larned et al. 2010). In addition, the first flood after drought abruptly relocates particulate P (Obermann et al. 2007), leaches P from stimulated mineralization processes in the sediment (Birch effect, Wilson and Baldwin 2008), and induces mineralization and leaching of terrestrial plant species that colonized the dry sediments and incorporated sedimentary P (Baldwin and Mitchell 2000). In addition, P from accumulated dead organic matter such as leaf litter is leached (Baldwin 1999, McComb et al. 2007). Figure 6.3 shows the amount of P that can be released from deposited leaf litter during inundation. The amounts tend to vary between the species with oak leaves releasing remarkably high amounts of P. The flood water may therefore have high P concentrations and high P loads. Re-flooded sediments may therefore be exposed to high P inputs. However, it is unclear how the precedent drying of the sediment can change its capability in retaining these inputs. Drying and rewetting cycles may play the key role in short-term and long-term control of P cycling in temporary aquatic systems, which is of great interest for management decisions.

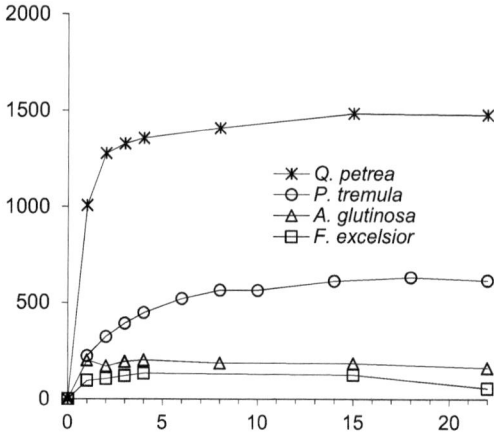

Figure 6.3 Cumulative release of soluble reactive phosphorus (P) from air-dried leaves of *Quercus petreae*, *Populus tremula*, *Alnus glutinosa*, and *Fraxinus excelsior*. Leaves from each species were incubated for a total of 22 days in separate water tanks containing artificial stream water (Fischer et al. 2006).

7. Effects of drying on phosphorus uptake in re-flooded lake sediments

This chapter includes a submitted article.

Dieter, D.; Herzog, C. & Hupfer, M. Effects of drying on phosphorus uptake in re-flooded lake sediments. submitted 2013.

7.1. Abstract

Sediment drying associated with large water table fluctuations is an increasingly common feature of temporary streams and lakes worldwide. Drying-induced sediment aeration and re-flooding periodically alter redox conditions and therefore stimulate redox-sensitive processes influencing phosphorus (P) dynamics. We experimentally tested the effects of drying on P dynamics and the P sorption potential by drying and re-flooding lake sediments in the laboratory. Wet and dried sediments were re-flooded with P-enriched water in sediment columns. Besides changes in P fractions, P uptake rates, and the pore water dynamics were measured in each column over the course of 36 weeks. Drying increased the fraction of stable P that was mobilized, stimulated the mineralization of organic P compounds, and increased the proportion of labile and reductant-soluble fractions. Drying furthermore reduced the P sorption affinity and capacity by up to 32% in batch equilibrium experiments, but also led to a 4-fold increase in sediment compaction which enhanced P uptake rates by a factor of 1.7 in sediment column experiments. Compaction due to drying also induced the development of a redoxcline below which P was mobilized. These results indicate that even a single drying event can result in the transformation of P components into more labile forms which accumulate in the uppermost sediment layer, therefore raising the potential for a pulsative P release under reducing conditions.

Effects of drying on phosphorus uptake in re-flooded lake sediments

7.2. Introduction

Most aquatic systems worldwide experience water level fluctuations, which expose previously flooded sediments not only to air but also to desiccation. Droughts can occur seasonally or as single events. Temporary streams are a common phenomenon (Tockner et al. 2009), but water level fluctuations in lakes and reservoirs also occur frequently (Wantzen et al. 2008b). Climate and land use changes, as well as enhanced water withdrawal, however, will increase the frequency and spatial extent of drying (Coops et al. 2003). Drying can range spatially from minor water level fluctuations affecting only the littoral zone, to large drawdown events which also affect the profundal zone of an aquatic ecosystem. Drying and re-flooding can have marked effects on biological and physical processes (Lake 2003), especially redox-sensitive processes such as carbon turnover and phosphorus (P) dynamics (Fabre 1988, de Vicente et al. 2010). The availability of P often controls the trophic state of freshwater ecosystems. The effects of droughts on P dynamics are therefore of particular relevance, but not yet clearly understood (Dahm et al. 2003, Withers and Jarvie 2008). Changes in P availability during drying and rewetting have been widely studied in forested soils, floodplains, and wetlands (e.g. Grierson et al. 1998, Baldwin and Mitchell 2000, Nguyen and Marschner 2005, Song et al. 2007, Zak and Gelbrecht 2007). For rivers and lakes, however, detailed information on P dynamics in sediments during desiccation and re-flooding is rare and mostly focuses on short-term P release. Most studies have reported an increase in P release following drying and re-flooding, due to bacterial mineralization (Qiu and McComb 1994, Watts 2000a, Xiao et al. 2012). In contrast to these studies, it has also been reported that P release after re-flooding may be restricted (Mitchell and Baldwin 1998) as bacteria experience an osmotic shock, and also since a shortage of sulfide in the oxidized sediments can prevent the release of P by iron reduction and iron fixation to sulfide (Caraco et al. 1989).

Detailed data on the underlying processes and changes in sediment sorption capacities are lacking or contradictory. Baldwin (1996) and Kerr et al. (2010), for instance, reported a reduction in P sorption affinity in dried sediments due to mineral aging, yet changes in P binding fractions were inconsistent. More detailed information would therefore be advantageous to evaluate possible effects of a perennial aquatic system when surface water disappears.

Re-flooding is often associated with an enhanced concentration of nutrients in the water (first flush) that are released from accumulated organic material, such as leaf litter and manure, or from microbial organic matter turnover in soils and sediments (Birch 1960, Watts 2000a+b, Qiu et al. 2004, Corstanje and Reddy 2004, McComb et al. 2007). In addition, P concentrations from point sources such as waste water inputs or from diffuse sources such as agricultural outputs are often higher during periods of water shortage or during first flood events (Withers and Jarvie 2008). Furthermore, remarkable amounts of P are also supplied to the sediments by seston sedimentation, especially in lakes (Hupfer et al. 1995, Kleeberg 2002). Information on the potential of dried and re-flooded sediments to retain these P inputs is lacking, and conclusions regarding the trophic conditions of affected ecosystems are unreliable. In addition, the P uptake capacity of sediments has often been estimated by using a microcosm batch equilibrium technique (Nair et al. 1984). However, this method disregards sediment bulk density and diffusion pathways, and the stratification of geochemical conditions such as redox gradients cannot be simulated. A comparison of P uptake in experimental sediment columns would therefore be more precise (Reddy et al. 1999).

The objective of this study was to investigate changes in the mobilization and uptake of P in iron-rich sediments exposed to drying and re-flooding with P-

Effects of drying on phosphorus uptake in re-flooded lake sediments

enriched water. We hypothesized that oxidative processes during drying would increase the fraction of labile P due to the mineralization of organic P, and that drying would reduce the P uptake potential of the sediments due to mineral aging.

7.3. Methods and Material

7.3.1. Sediment sampling and drying

In total, 17 undisturbed sediment cores (ø 6 cm) were collected with PVC tubes from the profundal zone of lake Müggelsee, a shallow polymictic and eutrophic lake southeast of Berlin, Germany (52°26'N, 13°39'E). The lake has an approximate area of 7.3 km^2, with a mean depth of 4.9 m. Limnological and sediment properties were previously described in detail (Driescher et al. 1993, Kleeberg and Kozerski 1997). Sediment cores had never been exposed to previous drying events. All 17 cores were sliced into 3 separate layers: upper layer (0 - 2.5 cm), lower layer (2.5 - 12.5 cm), and base layer (> 12.5 cm). Samples from each layer were sieved (< 2 mm) and pooled into three plastic containers. Subsamples of the upper and lower layers were air-dried in trays under laboratory conditions (20 °C) for 4 weeks to 3% residual gravimetric moisture, as observed in sediments of temporary waters (Kerr et al. 2010). Meanwhile, wet sediments of the two layers and the base layer were kept in closed containers at 4°C.

7.3.2. Batch equilibrium experiment

In order to determine sediment sorption affinity and sorption capacity (Reddy et al. 1999), linear, Langmuir, and Freundlich sorption models were fitted as appropriate to sorption curves. Sorption curves (i.e. sorbed P vs. equilibrium P concentrations) were determined in a batch equilibrium experiment (BEE) at 20 °C, following Nair et al. (1984). 0.5 g aliquots of dried or original wet sediments from both upper and lower layers were shaken for 3 h in 20 mL of a KH_2PO_4 solution with P concentrations ranging from 0 to 500 mg L^{-1} across 12 steps (Kleeberg et al. 2010). The slurry was centrifuged (10,000 g) and the supernatant was passed through 0.45 µm cellulose acetate syringe filters (Whatman) prior to the analysis of equilibrium soluble reactive P (SRP). SRP concentrations were determined with a spectrophotometer (San++ CFA, Skalar, Breda, Netherlands) following the molybdate-blue method (Murphy and Riley 1962). Prior to the addition of a P solution, SRP concentrations in the pore water of the wet sediment were determined by sediment centrifugation and photometric detection, and these values were used in P sorption calculations (Kleeberg et al. 2010).

7.3.3. Sediment column experiment

In order to estimate P uptake capacities with respect to the effects of diffusion and complex, long-term processes (Reddy et al. 1999), the three sediment layers were re-filled into 8 PVC tubes (ø 6 cm) to generate artificial sediment cores. All tubes were first loaded with 15 cm of the wet base material to allow for microbial re-colonization in the dried sediments. Half of the tubes (4) were stocked with dried sediment layers, and half with original wet sediment layers. The sediment dry mass of the dried upper and lower layers was equal to the sediment dry mass of the wet sediments of the same layers (determined at 105 °C for 8 h). The upper layer had a thickness of 2.5 cm in the wet sediments and 0.5 cm in the dried sediments, whereas the lower layer had a thickness of 10 cm in the wet sediments and 2.5 cm in the dried sediments. For all 8 artificial sediment columns, tubes were carefully flooded with tap water, put into large racks, and placed in a climate chamber at 10 °C. The overlying water was constantly aerated using flexible tubes, and was stocked with a KH_2PO_4 solution to obtain a P concentration of 2 mg L^{-1} so as to be greater than the P concentration of the original pore water (0.9 mg L^{-1} in the upper layer and 1.4 mg L^{-1} in the lower layer). Over a period of 36 weeks, sediment P uptake rates were determined by measuring SRP concentrations in the overlying water, after which concentrations would be returned to 2 mg L^{-1} by restocking. Water pH and specific conductivity were measured every 3 weeks with dipping probes (WTW, Weilheim, Germany). Two tubes with wet sediments and 2 tubes with dried sediments were previously equipped with rhizone pore water samplers (Eijkelkamp, Giesbeek, Netherlands). The pore water was sampled using 1 mL syringes at 5 dates and analyzed for (i) SRP and ferrous iron with a spectrophotometer (Sunrise, Tecan, Männedorf, Switzerland) using microtiter plates (Laskov et al. 2007), and (ii) sulfate with ion chromatography (LC-10A, Shimadzu, Kyoto, Japan).

At the end of the experiment, sediments from the 4 columns without rhizone pore water samplers were sliced into layers at 1 cm increments. For each layer, the samples of two columns were combined with either previously dried sediments or original wet sediments.

7.3.4. Sequential phosphorus fractionation

In order to determine changes in P binding fractions due to drying and P uptake, a sequential P fractionation was performed following Psenner et al. (1984) with modifications as described in Hupfer et al. (1995). This was performed on (i) the original wet and laboratory dried sediments of the upper and lower layers, and (ii) the layer samples from the sliced sediment columns at the end of the experiment. In the extracts, SRP and total P (TP) after $K_2S_2O_8$ (5%) digestion at 120°C were measured with a spectrophotometer (San++ CFA, Skalar, Breda, Netherlands). In addition, dissolved Fe^{2+} and Mn^{2+} concentrations in the bicarbonate/dithionite (BD) extract were measured with atom absorption spectrophotometry. TP concentrations in the sediments were determined by (i) direct photometric measurements after H_2SO_4/H_2O_2 digestion at 160°C (Kleeberg et al. 2010), and (ii) direct measurements after microwave-assisted aqua regia dissolution and detection with an ICP-OE spectrometer (iCAP 6000, Thermo Scientific, USA). The TP concentrations which were calculated as the sum of P fractions and by H_2SO_4/H_2O_2 digestion accounted for ≥ 83% and ≥ 94% of TP by ICP-OES detection, respectively. In addition to P, other elements (Al, Fe, Ca, Mn, and Mg) were also simultaneously extracted with aqua regia and measured by ICP-OES detection. Total carbon and nitrogen concentrations were detected with a C/N elemental analyzer (Elementar vario EL, Hanau, Germany).

All statistical analyses were performed using *R* software (version 2.14.2, Ihaka and Gentleman 1996) with the significance level set to $\alpha = 0.05$.

7.4. Results

7.4.1. Sediment properties

The collected sediment was characterized by the atomic ratios of Fe:S:P = 10:4:1 and H:C:N:S = 43:36:3:1. The loss on ignition accounted for 25% of total dry mass (DM). The sediment contained 130.2 mg Ca g^{-1} DM, 2.3 mg Mg g^{-1} DM, 64.0 mg Fe g^{-1} DM, 1.5 mg Mn g^{-1} DM, and 11.1 mg Al g^{-1} DM, and was therefore rich in iron.

7.4.2. Sediment drying

During drying in the laboratory, sediments showed cracks and compacted to 20% (upper layer, from 0.08 to 0.38 g DM cm^{-3}) and 25% (lower layer, from 0.10 to 0.40 g DM cm^{-3}) of the former thickness and did not re-expand during re-flooding. Drying also led to a shift in the proportion of P fractions in both layers (Table 7.1). Labile fractions increased (i.e. directly available P (NH_4Cl-P) and reductant-soluble P (BD-P)), while NaOH-SRP and organic P (NaOH-NRP) declined. The increase in NH_4Cl-P was on average four times higher than the amount of P that was taken up from the evaporating pore water (~ 11 µg P g^{-1}). The ratio of Fe:P in the BD extract increased from 2.5 to 2.8 in the upper layer and from 1.8 to 3.2 in the lower layer. More stable fractions (HCl-P and residual P) remained unaffected by drying and the TP yield in dried sediments was almost identical to that in wet sediments.

Table 7.1 Concentrations of phosphorus (P) fractions as determined by the sequential extraction of original wet and laboratory dried lake sediments taken from two different layers. DM, sediment dry mass, TP: total P, SRP: soluble reactive P, NRP: non-reactive P, BD: bicarbonate/dithionite. Values are means of laboratory duplicates.

		NH_4Cl	BD	NaOH			HCl	residual	sum
		TP	TP	SRP	NRP	TP	TP	TP	TP
				µg g^{-1} DM					
0-2.5 cm	wet	15	1788	725	506	1231	358	137	3529
	dried	73	2060	482	428	910	355	133	3530
2.5-12.5 cm	wet	4	1126	749	413	1162	393	129	2814
	dried	37	1364	608	342	950	377	129	2858

7.4.3. Batch equilibrium experiment

Our BEE demonstrated that drying causes a lower sorption potential. Equilibrium P concentrations at zero net sorption in the upper and lower layers were 6- and 3-times higher in dried sediments than in wet sediments (Table 7.2, EPC_0). Dried sediments also exhibited a slower increase in sorbed P with rising P concentrations (Table 7.2, k in all models). Both observations indicated a reduced sorption affinity of dried sediments. While the linear model fitted best for low P sorption (< 400 µg g^{-1}) in both treatments, Freundlich fitted badly for wet sediments. Higher P sorption values (> 500 µg g^{-1}) were thus selected, and a Langmuir sorption equation was fitted for positive sorption values (Table 7.2) (Barrow 1978, Reddy et al. 1999, Barrow 2008). The Langmuir model revealed a reduction in sorption capacities by 15% and 32% for dried sediments in the upper and lower layer, respectively (Table 7.2, q_{max}).

Table 7.2 Experimentally-determined sorption models for the uptake of phosphorus (P) in original wet and laboratory-dried lake sediments taken from two different layers. c: equilibrium P concentration in mg P L^{-1}, EPC$_0$: equilibrium P concentration at net zero sorption (model intercept with coordinate), q: µg adsorbed P per g sediment dry mass, q_0: intercept with ordinate, q_{max}: maximum sorption capacity in µg P g^{-1}, k: model coefficient denoting sorption affinity. Values are means of laboratory duplicates.

		linear q=k*c+q$_0$				Langmuir q=q$_{max}$*k*c/(1+c*k)			Freundlich q=k*c^n		
		for q < 400 µg P g^{-1}				for q > 0 µg P g^{-1}			for q > 500 µg P g^{-1}		
		EPC$_0$	k	q$_0$	R^2	q$_{max}$	k	R^2	k	n	R^2
0-2.5 cm	wet	0.04	19510	-703	0.993	7443	0.10	0.974	1948	0.25	0.971
	dried	0.25	713	-177	0.992	6304	0.03	0.955	584	0.40	0.997
2.5-12.5 cm	wet	0.02	39936	-787	0.908	10776	0.20	0.988	3453	0.24	0.883
	dried	0.07	2408	-168	0.969	7381	0.04	0.918	726	0.40	0.994

7.4.4. Sediment column experiment

The re-flooding of sediment columns with P-enriched water provided initial P uptake rates which were 1.7 times faster in previously dried sediments (55 ± 8 mg m^{-2} d^{-1}) than in original wet sediments (31 ± 6 mg m^{-2} d^{-1}) (Figure 7.1). The P uptake rates differed significantly between treatments (ANCOVA on log$_{10}$ transformed data with date as covariate: $F = 38.34$, $P < 0.001$) and declined more strongly and steadily in the dried sediment columns than in wet sediment columns. In the wet sediment columns, rates initially decreased but then increased again after 16 weeks, exceeding dried sediment rates after 20 weeks (7 ± 1 mg m^{-2} d^{-1} vs. 17 ± 6 mg m^{-2} d^{-1} by the end of the experiment). Replicates of previously dried sediments were very similar, whereas replicates of the wet sediments drifted apart slightly. The total quantity of P uptake during 36 weeks tended to be higher for previously dried sediments (4.1 ± 0.2 g m^{-2}) than for wet sediments (3.4 ± 0.8 g m^{-2}). The pH of the overlying water during P addition was higher for the previously dried sediments (8.6 ± 0.1, n = 48) than for the wet sediments (8.3 ± 0.2, n = 48) (t = 8.54, df = 88, P < 0.001). In addition, the specific conductivity was higher for the previously dried sediments (1096 ± 211 µS cm^{-1}, n = 40) than for the wet sediments (665 ± 102 µS cm^{-1}, n = 40) (t = 11.67, df = 56, P < 0.001).

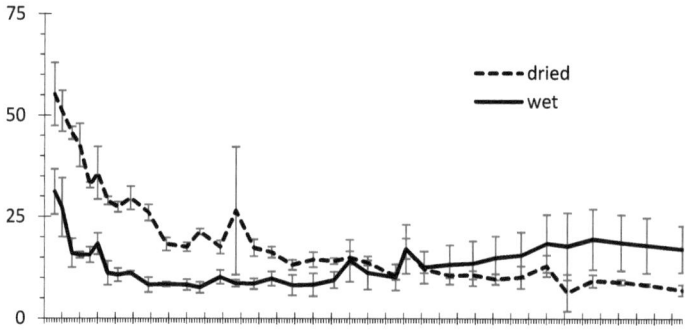

Figure 7.1 Rate of phosphorus (P) uptake (mean ± sdv, n = 4) in sediment columns flooded with water containing 2 mg P L^{-1}. Four columns contained original wet sediments and four columns contained previously dried sediments.

7.4.5. Sequential phosphorus fractionation

Changes in P fractions after P addition to previously dried or wet sediments were compared to the original material from both layers (Figure 7.2a, b, c). To isolate the processes caused by P addition from combined processes caused by drying and P addition, changes in the P fractions of dried sediments were calculated for both the original wet (Figure 7.2b) and the original dried sediments (Figure 7.2c). A comparison of these calculations revealed that the combined effects of drying and P addition were additive in the upper 2 cm, but were partly compensatory in deeper layers (NaOH and BD fractions). Since the outcome remains the same, the following descriptions of dried sediments refer only to the difference from the original dried sediments (Figure 7.2c). For both treatments, added P was mainly taken up by reductant-soluble compounds (BD fraction) (up to 1798 µg g^{-1} and 2480 µg g^{-1} in wet and dried sediments, Figure 7.2a, c), while P was lost from more stable P compounds (NaOH-SRP) (up to -354 µg g^{-1} and -315 µg g^{-1} in wet and dried sediments). The concentrations of directly available P (NH$_4$Cl-P) increased slightly (up to 68 µg g^{-1} in previously dried sediments) in both treatments. In wet sediments, total P uptake increased with depth down to 4.5 cm, below which changes in P fractions were negligible. In contrast, previously dried sediments took up P to a depth of only 2 cm, with a maximum uptake zone between 0.5 and 1 cm, where P uptake was higher than in the wet sediments. Below 2 cm, previously dried sediments experienced a net P loss. While more stable P compounds (NaOH-SRP) increased (up to 550 µg g^{-1}), P was lost from reductant-soluble compounds (BD-P) (up to -423 µg g^{-1}). In addition, P was lost from organic compounds (NaOH-NRP) (up to -244 µg g^{-1}) in all depths of previously dried sediments, while wet sediments gained organic P across all depths (up to 159 µg g^{-1}). The total net P uptake in the studied layers was greater than the total amount of P removed from the water column in both treatments (wet: 4.8 g m^{-2} vs. 3.4 g m^{-2}, dried: 6.9 g m^{-2} vs. 4.1 g m^{-2}).

TP calculated as the sum of sequentially extracted fractions was equivalent to 91% and 89% of the TP determined by hot H$_2$SO$_4$/H$_2$O$_2$ digestion and ICP-OES detection, respectively.

Effects of drying on phosphorus uptake in re-flooded lake sediments

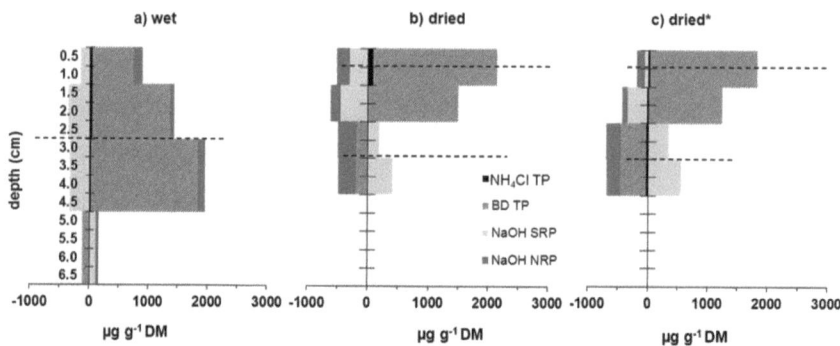

Figure 7.2 Changes in phosphorus (P) fractions as determined by sequential extraction of lake sediments that were re-flooded with P-enriched water (constantly restocked to 2 mg L^{-1}) over 36 weeks after having been dried or not dried (wet): a) & b) changes referring to the original wet sediment material, c) changes referring to dry sediment, isolating the effect of P addition. TP: total P, BD: bicarbonate/dithionite, SRP: soluble reactive P, NRP: non-reactive P, DM: dry mass of sediment. The dashed lines indicate the boundaries between the upper and lower layers (0 - 2.5 cm, 2.5 - 12.5 cm) and the wet base material (> 12.5 cm) that were originally sampled from the lake before homogenization, drying, and refilling of sediment columns. Values are means of laboratory duplicates obtained from pooled sediment samples. Fractions with negligible changes were excluded here (HCl-P, Residual-P).

With respect to the irreversible sediment compaction during drying, Figure 7.3 shows the absolute amounts of all P fractions at different depths by volume (i.e. P concentrations provided in µg cm^{-2} per 0.5 cm depth). This figure demonstrates the high accumulation of reductant-soluble P in the upper 0 - 1 cm layer due to compaction, and the high uptake of P in previously dried sediments (dried: 1.97 mg cm^{-3} vs. wet: 0.33 mg cm^{-3}).

Effects of drying on phosphorus uptake in re-flooded lake sediments

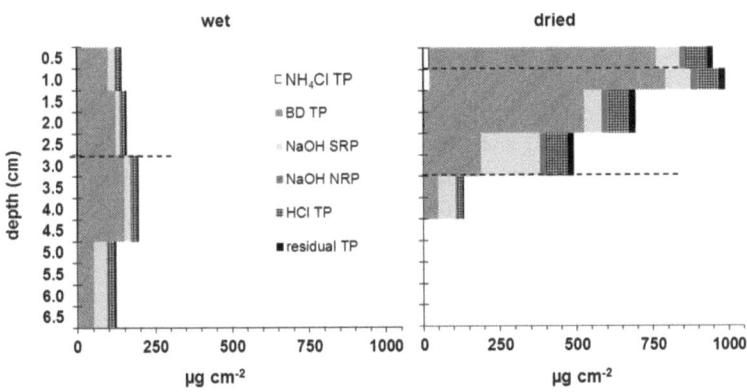

Figure 7.3 Fractions of phosphorus (P, given in μg cm^{-2} per 0.5 cm depth) determined by sequential extraction of lake sediments that were re-flooded with P-enriched water (constantly restocked to 2 mg L^{-1}) over 36 weeks after having been dried or not dried (wet). TP: total P, BD: bicarbonate/dithionite, SRP: soluble reactive P, NRP: non-reactive P. Dashed lines indicate the boundaries between the upper and lower layers (0 - 2.5 cm, 2.5 - 12.5 cm) and the wet base material (> 12.5 cm) that were originally sampled from the lake before homogenization, drying, and refilling of sediment columns. Values are means of laboratory duplicates obtained from pooled sediment samples.

Sediment drying raised reductant-soluble Fe and Mn concentrations (BD fraction), which increased further in the upper layer after re-flooding (Figure 7.4). Fe concentrations even increased down to a depth of 6 cm in the wet sediments, and therefore also expressed a relatively high Fe:P ratio (2.6 on average) in the BD fraction. Fe in the BD extract accounted for between 9 and 18% of the total Fe content of the wet sediment. In dried sediments, Fe concentrations rapidly declined below 1 cm depth and therefore showed relatively low Fe:P ratios approaching saturation (1.6 on average). The proportion of BD-Fe therefore rapidly declined from 16% down to 4% of total Fe. In both treatments, changes in BD-Fe were negatively correlated with changes in Fe extracted with HCl (r = -0.90, t = -4.56, df = 5, P < 0.01), especially for layers below a depth of 1 cm. Reductant-soluble Mn declined abruptly below 1 cm in both treatments. This was consistent with total Fe and Mn concentrations (Table 7.3).

Table 7.3 Concentrations of C, N, Fe, Mn, and P in lake sediments of two different layers (original) and after P enrichment of the same sediments that had previously been dried or not dried (wet). Values are means of laboratory duplicates obtained from pooled sediment samples.

		C	N	C:N	Fe	Mn	P
	cm	%	%	molar	‰		
original	0-2.5	16.6	1.5	11.3	64	1.5	3.8
	2.5-12.5	16.4	1.4	11.3	64	1.5	3.2
wet	0-1.0	15.6	1.4	11.4	68	2.6	4.4
	1.0-2.5	15.5	1.3	11.6	68	1.9	4.9
	2.5-4.5	16.0	1.5	11.0	69	1.2	4.8
	4.5-6.5	16.1	1.5	10.9	69	1.5	3.1
dried	0-1.0	15.5	1.3	11.5	70	2.0	5.5
	1-2	15.6	1.3	11.7	69	1.4	4.0
	2-3	15.7	1.3	11.7	71	1.5	2.9
	3-4	15.6	1.3	11.8	71	1.6	3.1

Carbon and nitrogen compounds were mineralized in the dried sediments and in the upper layer of the wet sediments as indicated by the decrease in %C and %N, which also led to a slight increase in the C:N ratio in these zones (Table 7.3).

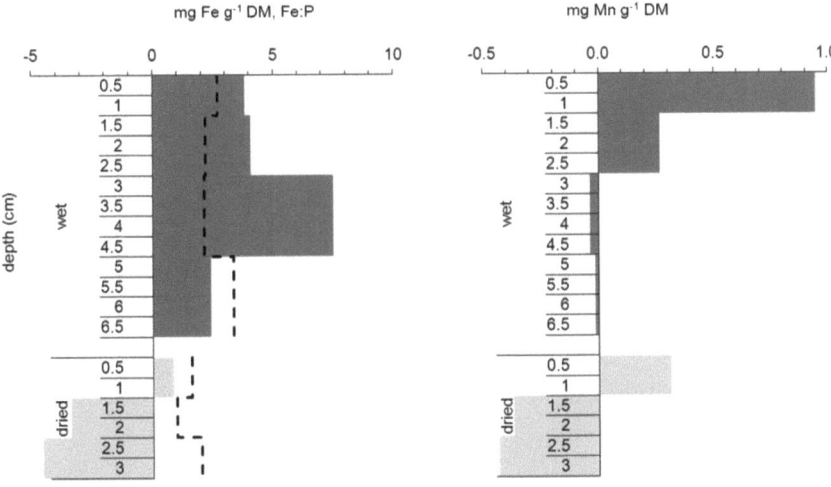

Figure 7.4 Changes in the concentrations of Fe^{2+} and Mn^{2+}, determined from a bicarbonate/dithionite (BD) extract during sequential fractionation of lake sediments that were re-flooded in sediment columns with P-enriched water over 36 weeks after having been dried or not dried (wet). Grey-shaded areas indicate changes in concentrations of Fe^{2+} and Mn^{2+}. The dashed line indicates the molar Fe:P ratio in the BD fraction at the end of re-flooding.

7.4.6. Pore water dynamics

In columns with wet and previously dried sediments, pore water P concentrations increased up to a maximum of 7 mg L^{-1} by the end of P addition (Figure 7.5). The pore water P in upper layers of previously dried sediments was many times greater than that of wet sediments, reaching maximum concentrations at a depth of 2 cm. In contrast, concentrations in wet sediments steadily increased with depth, reaching maximum concentrations at approximately 10 cm. Ferrous iron concentrations in wet sediments were initially negligible in the upper layer, then increased temporarily, but declined again at all depths towards the end of the experiment. In contrast, Fe^{2+} concentrations in previously dried sediments rapidly increased at depths of 0 - 4 cm, peaked at 2 cm with a maximum concentration of 50 mg L^{-1}, and later declined, leveling concentrations in the wet sediments. Sulfate concentrations in the pore water of the wet sediments declined steadily during the experiment, providing no differences between different depths. Previously dried sediments began with increased SO_4^{2-} concentrations of up to 1423 mg L^{-1} at a depth of 2 cm. This accounted for about 16.5% of the total S being oxidized.

After a lag time of ≥ 3 weeks these concentrations declined rapidly, leveling concentrations in wet sediments by the end of the experiment. Remarkably, sulfate concentrations decreased during week 8 in the same 2 cm. Sulfate reduction rates and ferrous iron mobilization/consumption rates were calculated by integrating concentrations across pore water depth profiles and the differences between pore water sampling dates. The development of the rates of sulfate and ferrous iron from one sampling to the following were positively correlated (r = 0.88, t = 5.74, df = 10, P < 0.001). The rates of both ions were highest for the previously dried sediments and for the interval between week 3 and 9 reaching maxima of 798 mg m^{-2} d^{-1} for sulfate and 15 mg m^{-2} d^{-1} for ferrous irion.

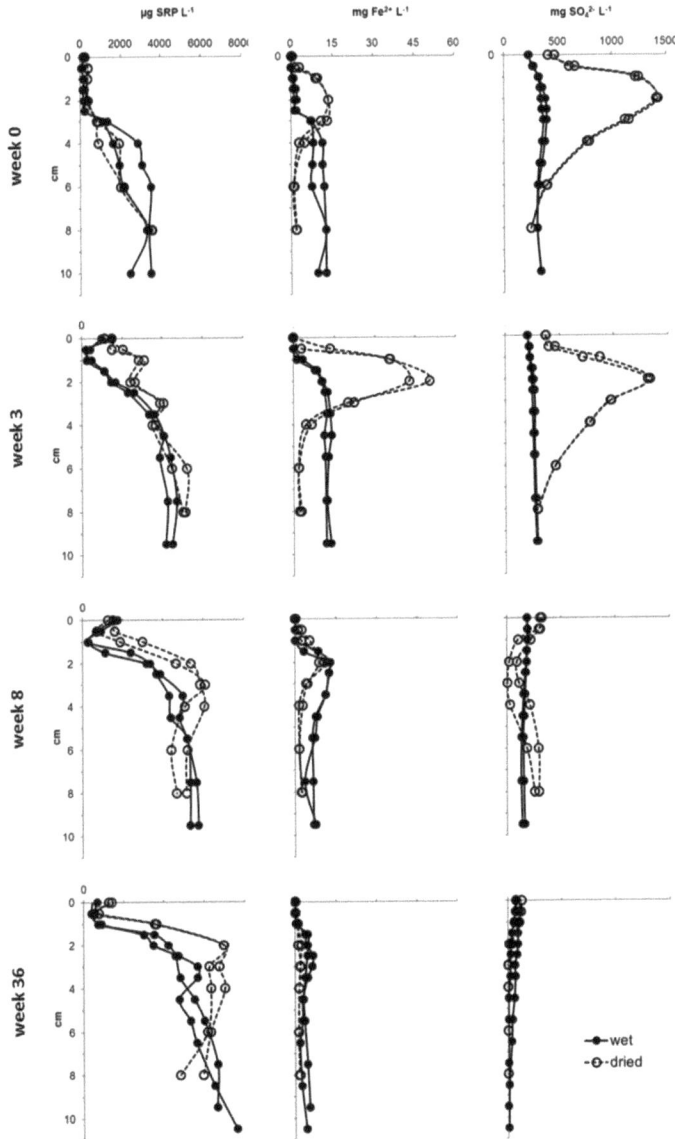

Figure 7.5 Vertical profiles for concentrations of soluble reactive phosphate (SRP), ferrous iron, and sulfate in the pore water of sediment columns over the course of 36 weeks of P addition to the overlying water column of previously dried and not dried (wet) lake sediments.

7.5. Discussion

7.5.1. Phosphorus binding fractions

The drying of previously non-dried lake sediments caused a shift in P binding fractions (Table 7.1). The amounts of reductant-soluble P and directly available P increased due to losses from more stable Fe and Al oxide-bound P, and the mineralization of organic P. The latter process was likely to be facilitated by the aeration of previously anoxic sediments (Qiu and McComb 1994) and was associated with a decrease in the C and N content of dried sediments (Table 7.3). The increase in directly available P could not be explained by the uptake of P from evaporating pore water alone. It was most likely of microbial origin, possibly released from lysed and disrupted cells under the stress of desiccation and osmotic shock of rapid re-flooding (Birch 1960, Baldwin and Mitchell 2000, Turner et al. 2003). The increase in reductant-soluble P was likely due to the oxidation of ferrous iron and the formation of Fe(III) oxyhydroxides, which preferentially bind P (Fox 1989), and would then be reduced again during BD extraction (Psenner et al. 1984). This shift in composition of P fractions was in accordance with the results from previous studies performed on dried river sediments (Kerr et al. 2010), reservoir sediments (Baldwin 1996), and marshland soils (de Groot and Fabre 1993).

7.5.2. Phosphorus sorption affinity and capacity

Drying also caused a decline in the P sorption affinity and capacity of sediments (Table 7.2), as has been observed in earlier studies (Baldwin 1996, Watts 2000a, de Vicente et al. 2010, Xiao et al. 2012), with sorption parameters very similar to the results of Kerr et al. (2010). However, we observed a six-fold increase in EPC_0, exceeding the results of Twinch (1987), where only a threefold increase of EPC_0 was observed. The decrease in P affinity may have been due to numerous causes. First, it may have been the result of the formation of sediment aggregates during compaction, which could have occluded potential sorption spaces in dried sediments. However, this contrasts with the fact that the compaction and building of aggregates did not influence the TP recovery by sequential extraction when compared to the wet sediments, suggesting that aggregates were totally resolved. Second, the aeration of anoxic wet sediments (e.g., by draining or shaking in water that contains oxygen) may have led to the oxidation of ferrous iron to amorphous Fe(III) hydroxides, which can easily bind P and therefore artificially raises the P sorption affinity. This would not explain, however, why fully oxidized dried sediments would have a lower sorption affinity. Third, elongated or repeated drying may have facilitated the formation of a more crystalline structure of amorphous Fe(III) hydroxides (aging), which involved the loss of potential sorption spaces for P, thereby reducing sorption affinity (Lijklema 1980, Baldwin et al. 2000, Qiu & McComb 2002, de Vicente et al. 2010). Fourth, the decomposition of organic matter could also have released labile P, increasing the EPC_0 measured upon re-flooding (Xiao et al. 2012). Finally, as mentioned above, drying causes microbial cell death (Turner and Haygarth 2001), which restricted microbial P uptake compared to the wet sediments. Kamp-Nielsen (1974) observed reduced P uptake from sterilized sediments and Watts (2000a) observed a reduced P release from un-sterilized sediments. Since our BEE was performed without fumigation, biological P uptake was enabled in addition to chemical P sorption.

Compared to wet sediments, the EPC_0 value for the dried upper sediment layer showed an extreme increase. It even exceeded long-term mean TP concentrations in the water of lake Müggelsee during summer (but not pore water concentrations). This implies that the dried sediments of lake Müggelsee might act as a major P source if drying and re-flooding occurred.

Effects of drying on phosphorus uptake in re-flooded lake sediments

7.5.3. Phosphorus uptake in sediment columns

The sediment column experiment partly confirmed and partly contradicted the results of the BEE. During the final third of the sediment column experiment, wet sediments experienced higher uptake rates than previously dried sediments, confirming the results of the BEE (Figure 7.1). As with previously dried sediments, uptake rates in wet sediments initially declined due to the rising saturation of P, yet uptake rates in wet sediments eventually increased again. An active population of microbiota may have contributed to biological P uptake processes in the sediments, as explained for the BEE. This is supported by the observed increase in organic P fractions in wet sediments, whereas organic P was lost in dried sediments. To corroborate this, phosphor-lipid fatty acid analysis was performed on residual sediment samples, revealing that the bacterial biomass was about 40% lower in previously dried sediments even after 8 months of re-flooding (Figure 7.7 in chapter 7.7 Appendix), although the base layer should have facilitated the microbial re-colonization of dried sediments. The structure of bacterial groups, however, did not differ between treatments, suggesting that the single drying treatment in this study did not select for specialized groups of bacteria (Baldwin and Mitchell 2000).

At the beginning of the sediment column experiment, P uptake rates in columns with previously dried sediments were almost double those in wet sediment columns, contrasting with the results of the BEE. This may be an effect of sediment compaction during drying. During the BEE, wet and dried sediment samples moved loosely through the liquid in shaken centrifuge tubes, thus the original bulk density was irrelevant. In contrast, the higher density of dried sediments in the columns provided a higher density of potential sorption places within shorter distances (Reddy et al. 1999), thus superimposing the effect of reduced sorption affinity. This was supported by the findings that P

Effects of drying on phosphorus uptake in re-flooded lake sediments

uptake in previously dried sediments was restricted to the upper 2 cm and was much higher there than in the wet sediments, where P also diffused into deeper layers. This may have led to higher uptake rates in previously dried sediments compared to wet sediments, although both sorption affinity and capacity were lower. These findings imply that results from nutrient sorption experiments involving the shaking of sediment must be interpreted carefully, taking into consideration *in situ* bulk density when accounting for the effects of drying and re-flooding. In other words, uptake rates cannot always be directly deduced from sorption affinity and capacity, as determined by a BEE.

Below the primary P uptake zone at 2 cm, processes appeared to be reversed, indicating a sharp redoxcline in previously dried sediment (Figure 7.6). In fact, reductant-soluble Fe and Mn expressed an abrupt decline below 1 cm in previously dried sediments (Figure 7.4). Within the top 1 cm, more Fe and Mn were in oxidized forms due to drying and water column aeration and could therefore be reduced and extracted with BD in addition to already reduced forms. Below that layer, anoxic conditions facilitated the reduction and fixation of Fe and Mn, thus fewer oxidized and reducible Fe and Mn forms were present. The reductant-soluble Mn increased in the uppermost sediment layer and then abruptly declined, indicating that it was mobilized in deeper layers, diffused upwards, and was finally precipitated at the oxygenated surface layer. However, a decrease in total Mn concentrations was not observed from the surface to 6.5 cm (Table 7.3), but it is possible that Mn was mobilized in even deeper layers.

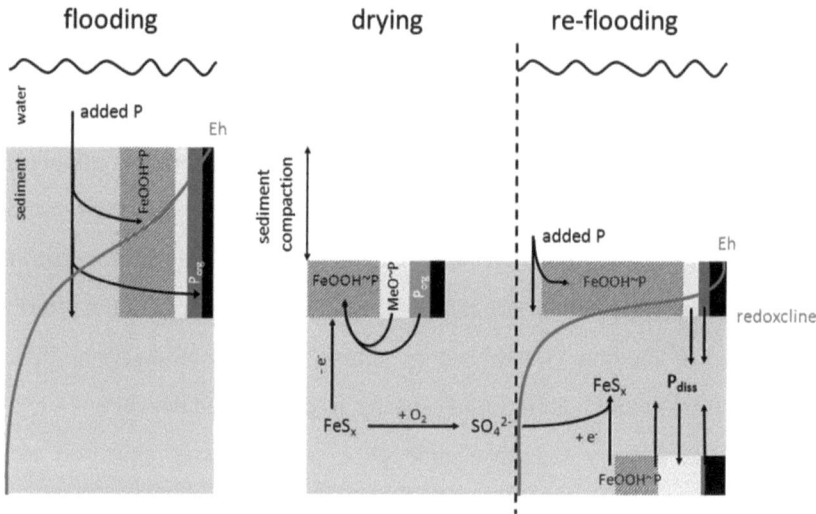

Figure 7.6 Simplified scheme of phosphorus (P) uptake and mobilization in lake sediments that are flooded with P-enriched water compared to sediments that are dried previous to re-flooding with P-enriched water. Shown dynamics are reduced to the most pronounced processes and changes in P pools with respect to the effect of drying. Stacked bars indicate the concentration of sequentially extractable P pools in the sediment at a particular layer. Eh: redox potential, P_{org}: organic P, MeO~P: NaOH-extractable P (mainly metal oxide-bound), P_{diss}: dissolved P.

In contrast, not only reducible Mn but also Fe increased in wet sediments even after re-flooding to a maximum depth of 6 cm. These may have been sequestered from the HCl fraction, yet the results also suggest that redox conditions did not foster Fe reduction and no such sharp redoxcline had developed in wet sediments. The redox potential likely declined smoothly with depth. The aeration of water columns in the laboratory and higher diffusion rates due to lower bulk densities may have supported higher redox levels in the wet sediment columns compared to original sediments in the lake or dried sediment columns. This was supported by the fact that almost no P was mobilized from the BD fraction in wet sediments, while it was lost in the anoxic layers of previously dried sediments. Here, this P fraction was fixed in more stable metal oxide compounds that were less sensitive to anoxic conditions as indicated by the increase in NaOH-SRP (Figure 7.2). In addition, much more P was mobilized from previously dried sediments, as indicated by the increasing P concentrations in the pore water (Figure 7.5). Remarkably, pore water P concentrations below 2 cm exceeded water column P concentration (2 mg L^{-1}) after only 3 weeks of re-flooding in both treatments. This would have restricted diffusive P transport to deeper layers (> 2 cm). Despite this, in the wet sediments we observed maximal P uptake between 2.5 and 4.5 cm. It was unlikely to be taken up prior to the increase in pore water P, since the amount of P taken up in this layer far exceeded the amount of P removed from the water column within the first 3 weeks. A high P uptake in this layer was more likely due to the corresponding increase in reductant-soluble Fe in the same layer (Figure 7.4). The additional P must have been mobilized in deeper layers, diffused upwards, and sorbed to iron compounds as indicated by the increase in reductant-soluble P (Figure 7.2) and the slope of the pore water P profile (Figure 7.5). This is corroborated by the surplus of P uptake detected by sequential fractionation compared to the total amount of P removed from the water column.

The development of P, Fe^{2+}, and SO_4^{2-} concentrations in the pore water also demonstrated the change between aerobic and anaerobic processes (Figure 7.5): During sediment preparation (sieving and mixing) and especially during sediment drying, oxidative processes resulted in the formation of ferric iron and sulfate. Hence, high amounts of sulfate were released into the pore water and overlying water after re-flooding, which explained the increase in conductivity in those columns (Baldwin and Mitchell 2000). With the onset of oxygen depletion after re-flooding, ferric iron was reduced again as indicated by the increase of ferrous iron in the pore water. However, with the onset of sulfate respiration at further decreasing redox levels, sulfate and ferrous iron concentrations declined, likely due to the formation of sulfide, precipitating ferrous iron (FeS$_x$, Figure 7.6) (Baldwin et al. 2000). This indicated that obligate anaerobic bacteria were, unexpectedly, not killed or suppressed by the desiccation (Mitchell and Baldwin 1998, Schönbrunner et al. 2012). Processes were highly active at a depth of 2 cm where ferrous iron concentrations temporarily peaked and sulfate was rapidly depleted (Figure 7.5). However, sulfate reduction rates were within the range of other lakes (Holmer and Storkholm 2001). The high sulfate consumption may be due to the fact that previous sediment drying stimulated aerobic metabolism, resulting in a high oxygen demand upon rewetting (Schönbrunner et al. 2012). This led to a rapid oxygen depletion at depth of 1 - 2 cm, while the layers above received oxygen from the aerated supernatant water. Sulfate and ferric iron were rapidly reduced at the same depth as high P uptake, resulting in low Fe:P ratios in the BD extracts approximating saturation in that depth. In anoxic layers below 2 cm of the previously dried sediments, a net loss of TP from the sediments was observed, raising pore water P concentrations. The fixation of iron in sulfides mobilized P and decoupled the Fe-P-cycle in dried sediments. The mobilized P most likely diffused upwards and was retained at the lower edge of the redoxcline as indicated by high TP concentrations at 1 – 2 cm immediately below the supposed redoxcline, and below the peak depth of TP increase at 0.5 – 1 cm (Figure 7.2). The increase in TP concentrations in upper layers with simultaneous TP decrease in deeper layers was also described by de Groot and van Wijck (1993), who also attributed this to an upward flux of P.

We detected no drying-related changes in HCl-extractable P (Table 7.1), whereas Baldwin (1996) reported an increase in dried sediments. Kerr et al. (2010) detected inconsistent changes in this fraction depending on sediment characteristics. This suggests that the form and amplitude of sediment reactions to drying and re-flooding are dependent on the native composition of the sediment. We used sediments which had a high Fe:P ratio and high S content. Most P was therefore bound to iron and was highly sensitive to changes in redox conditions with a high potential for sulfate-mediated P mobilization. In iron-poor or more calcareous sediments the Ca-bound P may be differently affected. Studies comparing different sediment types are necessary to elucidate the particular effects associated with each type. This would be important to deduce possible management implications for various aquatic ecosystems. De Groot and van Wijck (1993) proposed repeated drying and re-flooding as a management tool to take advantage of higher P release to flush out sediment P. Our results show that inundation after drying may indeed lead to a mobilization of P under anaerobic conditions, but P release might be restricted by repeated drying and rewetting (Wilson and Baldwin 2008). Drawdown does not seem to be a long-term option for the removal of P from the water body or retention in the sediments. If we apply the mean P uptake rates at the end of our experiment and presume that they are roughly stable, it would take another 10 weeks for the wet sediments to reach the same total amount of retained P as the dried sediments. In other words, within 46 weeks (almost one year), both systems would be able to retain the same amount of P. However, the dynamics and structure of the sediments would be different, with undisturbed wet sediments apparently being more efficient at long-term P uptake. Another proposition for management was to remove the degraded upper layer as critically discussed by Kleeberg and Kozerski (1997), Kleeberg and Kohl (1999), Watts (2000a), and Zak and Gelbrecht (2007) for reservoirs and re-wetted fens. The release of P and other substances upon re-flooding would be restricted, the function to retain P would be preserved, and the removed material could be transferred to agricultural sites to serve as a natural fertilizer. Total lake drawdown is, however, an extreme case of sediment exposition. Many lakes exhibit regular water level fluctuations that affect only part of the ecosystem. It is therefore difficult to predict the effects of littoral drying on the whole ecosystem. The extent to which a system is affected strongly depends on the morphology, with shallow lakes such as lake Müggelsee, or lakes with broad and shallow littoral zones, being the most sensitive (Wantzen et al. 2008a). In addition, consequences for rivers are distinct, because they are more affected by longitudinal flow and are therefore characterized by shorter residence times, higher turnover rates, and transportation resulting in a more intense aquatic-terrestrial link (Wantzen et al. 2008a), favoring for instance the input of nutrients such as P.

This study emphasizes that even one single drying event can already markedly influence processes involving the transformation of P species and functions of uptake and mobilization, which are primarily driven by oxidation and sediment compaction. When perennial systems with similar sediment characteristics as in this study dry out, P will accumulate in the top layer, representing an important P source after re-flooding.

7.5.4. Conclusion

The drying of formerly inundated sediments reduces P sorption affinity and capacity, stimulates mineralization of organic P compounds, results in a shift towards reductant-soluble, iron-bound P, and leads to an irreversible compaction of the sediments (Figure 7.6). Therefore, P uptake characteristics should not be deduced solely from small-scale experiments involving sediment shaking. Processes upon re-flooding with P-enriched water facilitate the formation of a sharp redoxcline, promoting the mobilization of P in the anoxic layer due to organic P mineralization and decoupling of the Fe-P-cycle. A single drying and re-flooding event can result in the accumulation of reductant-soluble P in the near surface sediment layers, thus raising the potential for P release into the water body under reducing conditions.

7.6. Acknowledgements

The authors like to thank S. Schiller, S. Jordan, and B. Schütze for their help with sampling and experimental set-up. Sincere thanks are also given to S. Schell, G. Siegert, H.-J. Exner, K. Premke, and A. Krüger for analytical support. Critical comments by A. Kleeberg and J. Lewandowski improved the manuscript. We thank Soren Brothers for professional language corrections.

7.7. Appendix

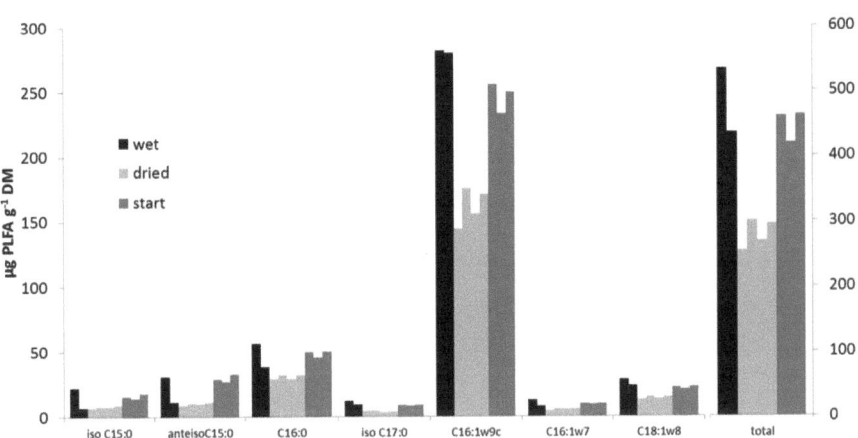

Figure 7.7 Phospho-lipid fatty acid (PLFA) extracted from sediments (0 - 1 cm) of lake Müggelsee (start) and after having been dried or not dried (wet) and re-flooded with P enriched water for 36 weeks. PLFA were extracted following the procedure of Steger et al. (2011) using gas chromatography and mass spectrometry detection.

8. Synthesis

The objectives of this dissertation were to assess to what extent a dry period preceding a wet period in aquatic ecosystems would influence leaf litter decomposition, as a critical process of carbon turnover, and the mobilization and uptake dynamics of phosphorus (P) in sediments, as a critical nutrient determining biological productivity.

The decomposition of organic matter is the key player in the global carbon cycle, which is predominantly discussed in the climate change debate. It involves the release of relevant greenhouse gases such as carbon oxides and methane. The decomposition of organic material from fossil resources and long-term storage in peat and permafrost soils is widely discussed. However, short-term cycles cannot be disregarded as the largest natural fluxes. It was estimated that the decomposition of leaf litter and fine root material accounts for about 70% of the total annual carbon flux (68 Pg C per year, Raich and Schlesinger 1992). Although terrestrial decomposition rates are expected to increase in response to the predicted global climate change (Tuomi et al. 2009), decomposition rates may be reduced in regions with increasing aridity, especially in temporary aquatic systems. However, climate change will not only directly affect decomposition rates by changes of temperature and precipitation. Climate change will also result in shifts in the community structure of plants worldwide, therefore affecting the chemical quality and decomposability of the shed litter (e.g. coniferous versus foliage litter, chapter 3). In addition climate change will also change the production rate of litter, for example due to the alteration of precipitation and sunshine hours or shorter frost periods. Future trends are therefore hard to model or to predict without a detailed understanding of the underlying processes. The effects of critical events such as droughts and floods need to be studied more in detail.

Synthesis

The studies presented in this book revealed that the disappearance of surface water flow affected leaf litter decomposition processes in streams (chapters 4, 5). Photodegradation of leaves as induced by solar radiation on dry streambeds and anaerobic fermentation as occurring in anoxic ponds enhanced the leaching of nutrients and labile carbon compounds from leaf litter. These preconditioning processes reduced the leaf quality considering it a substrate for decomposer communities such as macroinvertebrates and microorganisms. In fact, living fungal biomass was repressed and a change in the fungal community structure was observed. As a result, leaf decomposition rates in flowing water were reduced for preconditioned leaves, which held true for a range of streams and leaf species differing in chemical characteristics and quality. The results suggest that in streams developing seasonal flow intermittence or more severe droughts, preconditioning will influence leaf litter processing towards lower rates of microbially-mediated turnover and towards poorer quality of downstream-transported material.

The preconditioning by leaching does not only degrade accumulated leaf litter, but also represents an important source of nutrients for aquatic communities upon rewetting (Figure 6.2). Increased nutrient concentrations are therefore observed in so-called first flush events (McComb et al. 2007). It is however largely unclear, how these nutrients can be taken up in sediments or stimulate and inhibit aquatic organisms. Especially P is commonly regarded as the critical nutrient that limits biological productivity in terrestrial and aquatic ecosystems. Therefore, the effects of critical events such as droughts and floods on P dynamics need further research. The study presented in this book revealed that the drying of sediments affected the mobility of P (chapter 7). Drying mobilized more stable P fractions, stimulated the mineralization of organic P compounds, and caused a shift towards the proportion of labile and reductant-soluble, iron-bound P fractions. Drying reduced the P sorption

Synthesis

affinity and sorption capacity of the sediment, but also led to a sediment compaction, which in contrast enhanced initial P uptake rates after re-flooding. Compaction due to drying also induced the development of a sharp redoxcline that fostered P mobilization in deeper layers. The results indicated that a single drying event can result in the decoupling of the iron-phosphorus-cycle and a transformation of P components into more labile forms, which are accumulated in the near-surface sediment layer. This might raise the potential of high P release under reducing conditions after re-flooding.

As expected, the studies showed that temporary aquatic zones are characterized by highly dynamic ecosystem functioning compared to permanent systems. Therefore it seems appropriate to describe these systems with a combination of concepts that describe continuous (river continuum concept, Vannote et al. 1980) as well as pulsed processes (flood pulse concept, Junk et al. 1989). Larned et al. (2010) proposed new conceptual models for temporary streams that describe rivers and streams as *longitudinal, punctuated biochemical reactors*, meaning that continuous process (during wet periods) are interrupted from time to time (by dry periods). This interruption interferes with longitudinal gradients and introduces other processing modes (e.g. terrestrial), which results in processing cycles determined by drying/re-reflooding and transport/deposition. This may be comparable to the spiraling concept (Newbold et al. 1981). One spiral would be determined by one cycle of drying and re-flooding until the next drying event.

The biota in systems with periodic water fluctuations was repeatedly shown to be well adapted and specialized by either resistance or resilience. Insects were found to be most efficient in coping with seasonal droughts (Lake 2011, chapter 4) since they emerge at adult stages. However, during re-colonization after re-flooding all adapted biota rely on the availability of substrates. Detritus and litter that was deposited during the dry period therefore represents a

Synthesis

major food source, although it can be degraded and thus of lower quality (chapters 4 and 5). Riparian vegetation should therefore be more protected and afforestation should include careful species selection. However, a shift in the community structure of vascular plant species is generally expected along with global climate change, which itself will lead to changes in the quality of leaf litter and its leachates.

The periodic water level fluctuations represent disturbances for aquatic as well as terrestrial communities. It was proposed that the intermediate disturbance hypothesis (Connell 1978) could apply to these communities (Lake 2011). A high biotic diversity is the result of two or more communities that find their temporal niche in either the aquatic, terrestrial, or transitional period. Each community itself may have a higher diversity due to the fact that periodic disturbances limit the dominance of single species. In addition to species diversity, the intermediate disturbance hypothesis may also apply to the diversity of functional ecosystem processes. Drying, desiccation, flood pulse, and inundation may create a temporal diversity of processes that contribute to nutrient and carbon cycling. Temporary systems linking aquatic and terrestrial phases may therefore represent hot spots that also experience hot moments (McClain et al. 2003). However, the occurrence of severe and unpredictable (supra-seasonal) events may drastically alter the structure of communities and ecosystem processes resulting in an abrupt decline in diversity. This is, however, not well documented yet (Lake 2011). It remains, for instance, largely unrevealed whether there are critical thresholds in the duration, spatial extent, and severity of the periods that may lead to fundamental shifts in the diversity of species and processes (Steward et al. 2012). Resistance and resilience capacities are determining the thresholds. They control to what extent an ecosystem is deflected by a disturbance and how it can recover and return to the former state. Fisher et al. (2003) proposed the concept of a telescoping

Synthesis

ecosystem comparing nested subsystems of an aquatic system (central channel, hyporheic and parafluvial sediments, riparian zone) with segments of a telescope. These segments elongate and retract by disturbance and postdisturbance recovery, with the amplitude and rate being determined by the subsystems resistance and resilience capacity. Another option would be to compare the elongation and retraction of the telescope segments with the spatial expansion and contraction of the aquatic subsystems along with drying and re-flooding (Stanley et al. 1997).

Single events can markedly affect annual fluxes and budgets making it highly important to put more emphasis on studying and monitoring extreme events before the performance of whole catchment calculations or even the deduction of concepts and models. Concepts and management tools developed for permanent systems cannot be directly transferred to fluctuating systems. For instance, the original reference conditions suggested in the EU-WFD are indeed mostly not applicable for temporary systems. The studies presented in this book and all studies involved in the EU-project MIRAGE provided important background information on the timing and amplitude of ecosystem processes to define reference conditions and to suggest monitoring and management strategies. These results could successfully contribute to the setup of the MIRAGE toolbox, which represents an integrated assessment tool for temporary streams (Prat et al., submitted). The included tools encompass (i) the determination of the hydrological regime (e.g. permanent or temporary), (ii) protocols for biological and chemical sampling, (iii) the criteria of reference conditions, (iv) the assessment of hydrological modifications, and (v) protocols to measure the ecological status (including structural and functional methods). Tools are supposed to be applied sequentially to determine the ecological and chemical status of a temporary stream in order to deduce management strategies. However, droughts are global phenomena and must therefore be

Synthesis

incorporated not only into the EU-WFD but into water management plans worldwide (Humphries and Baldwin 2003).

The EU-WFD aimed to achieve a *good status* for European rivers by 2015. The report from the European Commission released by the end of 2012 shows that almost all countries have developed and approved river basin management plans (RBMP) (European Commission 2012). However, only a few RBMP were approved in Spain, while Portugal and Greece have not approved any RBMP. It is striking that the delay of EU-WFD implementation concerns countries where temporary aquatic systems typically occur. The good status will not be achieved by 2015 so that the deadline was extended until 2027 at the earliest.

During the last decades more emphasis has been put on studying the effects of droughts in aquatic ecosystems (Lake 2011). A lot of information has been collected on the effects on aquatic biota such as fish, macroinvertebrates, and algae. Little attention has been paid to microbes (e.g. bacteria and fungi) - an important part of the food web and essential link in the global carbon cycle. The knowledge about the influence of droughts on processes is very limited. More research is needed to understand its impact on ecosystem functioning and to assess global fluxes of elements. The effects on estuaries and oceans are also widely unknown. This information is crucial for the development of management strategies and policies such as the EU-WFD. It becomes even more relevant since ecosystems are facing increasing threats that go along with global change. In addition, the fluctuation of drying and re-flooding has been mostly studied and interpreted from a single perspective analyzing droughts and floods as disturbances from either the aquatic or the terrestrial point of view. Future research studying the aquatic-terrestrial transition zone (Junk et al. 1989) should take into account that water level fluctuations combine both. To approach this, it is necessary to study not only the effects of the event per se, but also how it influences the following period when water flow recovers or floodplains dry up. So far, it has remained uncertain which effects persist and for how long and which effects disappear so that the system could return to its former state. Hysteresis effects may play a significant role (chapter 7). In addition, droughts usually occur gradually, while flooding is often referred to as an abrupt event (Lake 2011). However, little information is available on the time-scale of the aquatic-terrestrial transition and the distinction between the effects of abrupt and gradual events. The studies presented in this dissertation attempted to head towards this future direction by studying the effects of a drought on the following wet period and underpin that this concept should be followed in future research.

9. References

Acuña, V.; Giorgi, A.; Munoz, I.; Sabater, F. & Sabater, S. **2007** Meteorological and riparian influences on organic matter dynamics in a forested Mediterranean stream. *Journal of The North American Benthological Society* 26:54-69.

Acuña, V.; Muñoz, I.; Giorgi, A.; Omella, M.; Sabater, F. & Sabater, S. **2005** Drought and postdrought recovery cycles in an intermittent Mediterranean stream: structural and functional aspects. *Journal of The North American Benthological Society* 24:919-933.

Albariño, R.; Villanueva, V. & Canhoto, C. **2008** The effect of sunlight on leaf litter quality reduces growth of the shredder Klapopteryx kuscheli. *Freshwater Biology* 53:1881-1889.

Albrectsen, B. R.; Bjorken, L.; Varad, A.; Hagner, A.; Wedin, M.; Karlsson, J. & Jansson, S. **2010** Endophytic fungi in European aspen (*Populus tremula*) leaves-diversity, detection, and a suggested correlation with herbivory resistance. *Fungal Diversity* 41:17-28.

Alcorlo, P.; Díaz, P.; Lacalle, J.; Baltanás, A.; Florín, M.; Guerrero, M. & Montes, C. **1997** Sediment features, primary producers and food web structure in two shallow temporary lakes (Monegros, Spain). *Water, Air, and Soil Pollution* 99:681-688.

Allan, J. D. & Castillo, M. M. **2007** Stream Ecology - Structure and function of running waters. *Springer, Dordrecht, Netherlands,* 436pp.

Anesio, A.; Denward, C.; Tranvik, L. & Graneli, W. **1999** Decreased bacterial growth on vascular plant detritus due to photochemical modification. *Aquatic Microbial Ecology* 17:159-165.

References

Artigas, J.; Gaudes, A.; Munoz, I.; Romaní, A. M. & Sabater, S. **2011** Fungal and bacterial colonization of submerged leaf litter in a Mediterranean stream. *International Review of Hydrobiology* 96:221-234.

Austin, A. T. & Ballaré, C. L. **2010** Dual role of lignin in plant litter decomposition in terrestrial ecosystems. *Proceedings of the National Academy of Sciences of the United States of America - PNAS* 107:4618-4622.

Austin, A. T. & Vivanco, L. **2006** Plant litter decomposition in a semi-arid ecosystem controlled by photodegradation. *Nature* 442:555-558.

Baldwin, D. S. & Mitchell, A. M. **2000** The effects of drying and re-flooding on the sediment and soil nutrient dynamics of lowland river-floodplain systems: A synthesis. *Regulated Rivers-Research & Management* 16:457-467.

Baldwin, D. S. **1996** Effects of exposure to air and subsequent drying on the phosphate sorption characteristics of sediments from a eutrophic reservoir. *Limnology and Oceanography* 41:1725-1732.

Baldwin, D. S. **1999** Dissolved organic matter and phosphorus leached from fresh and `terrestrially' aged river red gum leaves: implications for assessing river-floodplain interactions. *Freshwater Biology* 41:675-685.

Baldwin, D. S.; Gigney, H.; Wilson, J. S.; Watson, G. & Boulding, A. N. **2008** Drivers of water quality in a large water storage reservoir during a period of extreme drawdown. *Water Research* 42:4711-4724.

Baldwin, D. S.; Mitchell, A. M. & Rees, G. N. **2000** The effects of in situ drying on sediment-phosphate interactions in sediments from an old wetland. *Hydrobiologia* 431:3-12.

Baldy, V.; Chauvet, E.; Charcosset, J. Y. & Gessner, M. O. **2002** Microbial dynamics associated with leaves decomposing in the mainstem and floodplain pond of a large river. *Aquatic Microbial Ecology* 28:25-36.

References

Baldy, V.; Gessner, M. O. & Chauvet, E. **1995** Bacteria, fungi and the breakdown of leaf-litter in a large river. *Oikos* 74:93-102.

Bärlocher, F. & Graça, M. A. S. **2005** Total phenolics. In: Graça, M. A. S.; Bärlocher, F. & Gessner, M. O. (eds.) Methods to study litter decomposition: A practical guide. *Springer, Dordrecht, Netherlands,* 97-100.

Bärlocher, F. **2005** Leaf mass loss estimated by litter bag technique. In: Graça, M. A. S.; Bärlocher, F. & Gessner, M. O. (eds.) Methods to study litter decomposition: A practical guide. *Springer, Dordrecht, Netherlands,* 37-42.

Bärlocher, F.; Mackay, R. J. & & Wiggins, G. B. **1978** Detritus processing in a temporary vernal pool in southern Ontario, Canada. *Archiv für Hydrobiologie* 81:269-295.

Barrow, N. J. **1978** Description of phosphate adsorption curves. *Journal of Soil Science* 29:447-462.

Barrow, N. J. **2008** The description of sorption curves. *European Journal of Soil Science* 59:900-910.

Bates, B. C.; Kundzewicz, Z. W.; Wu, S.; Palutikof, J. P. (eds.) **2008** Climate change and water. Technical paper VI of the Intergovernmental Panel on Climate Change. *WMO, UNEP, IPCC Secretariat, Geneva, Switzerland,* 210pp.

Battle, J. M. & Golladay, S. W. **2001** Hydroperiod influence on breakdown of leaf litter in cypress-gum wetlands. *American Midland Naturalist* 146:128-145.

Benfield, E. F. **2006** Decomposition of leaf material. In: Hauer, F. R. & Lamberti, G. A. (eds.) Methods in stream ecology. 2nd edn. *Elsevier Academic Press, Burlington, USA,* 711-720.

References

Beniston, M.; Stephenson, D. B.; Christensen, O. B.; Ferro, C. A. T.; Frei, C.; Goyette, S.; Halsnaes, K.; Holt, T.; Jylhä, K.; Koffi, B.; et al. **2007** Future extreme events in European climate: an exploration of regional climate model projections *Climatic Change* 81:71-95.

Birch, H. F. **1960** Nitrification in soils after different periods of dryness. *Plant and Soil* XII:81-96.

Bloesch, J. & Uehlinger, U. **1990** Epilimnetic carbon flux and turnover of different particle-size classes in oligo-mesotrophic lake Lucerne, Switzerland. *Archiv fürHydrobiologie* 118:403-419.

Bond, N. R.; Lake, P. S. & Arthington, A. H. **2008** The impacts of drought on freshwater ecosystems: an Australian perspective. *Hydrobiologia* 600:3-16.

Boström, B. & Pettersson, K. **1982** Different patterns of phosphorus release from lake-sediments in laboratory experiments. *Hydrobiologia* 91:415-429.

Boström, B.; Andersen, J. M.; Fleischer, S. & Jansson, M. **1988** Exchange of phosphorus across the sediment - water interface. *Hydrobiologia* 170:229-244.

Boulton, A. J. & Boon, P. I. **1991** A review of methodology used to measure leaf litter decomposition in lotic environments - time to turn over an old leaf. *Australian Journal of Marine and Freshwater Research* 42:1-43.

Boulton, A. J. & Lake, P. S. **1990** The ecology of 2 intermittent streams in Victoria, Australia .1. Multivariate analyses of physicochemical features. *Freshwater Biology* 24:123-141.

Boulton, A. J. & Lake, P. S. **1992** Benthic organic-matter and detritivorous macroinvertebrates in 2 intermittent streams in south-eastern Australia. *Hydrobiologia* 241:107-118.

References

Boulton, A. J. **1991** Eucalypt leaf decomposition in an intermittent-stream in south-eastern Australia. *Hydrobiologia* 211:123-136.

Brandt, L. A.; Bohnet, C. & King, J. Y. **2009** Photochemically induced carbon dioxide production as a mechanism for carbon loss from plant litter in arid ecosystems. *Journal of Geophysical Research-Biogeosciences* 114:G02004, doi:10.1029/2008JG000772.

Bridgham, S. D. & Richardson, C. J. **2003** Endogenous versus exogenous nutrient control over decomposition and mineralization in North Carolina peatlands. *Biogeochemistry* 65:151-178.

Canhoto, C. & Laranjeira, C. **2007** Leachates of *Eucalyptus globulus* in intermittent streams affect water parameters and invertebrates. *International Review of Hydrobiology* 92:173-182.

Caraco, N. F.; Cole, J. J. & Likens, G. E. **1989** Evidence for sulfate-controlled phosphorus release from sediments of aquatic systems. *Nature* 341:316-318.

Casamayor, E. O.; Massana, R.; Benlloch, S.; Øvreås, L.; Díez, B.; Goddard, V. J.; Gasol, J. M.; Joint, I.; Rodríguez-Valera, F. & Pedrós-Alió, C. **2002** Changes in archaeal, bacterial and eukaryal assemblages along a salinity gradient by comparison of genetic fingerprinting methods in a multipond solar saltern. *Environmental Microbiology* 4:338-348.

Casas, J. J. & Gessner, M. O. **1999** Leaf litter breakdown in a Mediterranean stream characterised by travertine precipitation. *Freshwater Biology* 41:781-793.

Compton, J. S.; Mallinson, D. J.; Glenn, C. R.; Filippelli, G.; Follmi, K.; Shields, G. & Zanin, Y. **2000** Variations in the global phosphorus cycle. In: Glenn, C. R. (ed.) Marine authigenesis: From global to microbial. *Society for Sedimentary Geology (SEPM) special publication.* Tulsa, Oklahoma, USA, 21-33.

References

Connell, J. H. **1978** Diversity in tropical rain forests and coral reefs. *Science* 199:1302-1310.

Coops, H.; Beklioglu, M. & Crisman, T. L. **2003** The role of water-level fluctuations in shallow lake ecosystems - Workshop conclusions. *Hydrobiologia* 506-509:23-27.

Corstanje, R. & Reddy, K. R. **2004** Response of biogeochemical indicators to a drawdown and subsequent reflood. *Journal of Environmental Quality* 33:2357-2366.

Corti, R. & Datry, T. **2012** Invertebrates and sestonic matter in an advancing wetted front travelling down a dry river bed (Albarine, France). *Freshwater Science* 31:1187-1201.

Dahm, C. N.; Baker, M. A.; Moore, D. I. & Thibault, J. R. **2003** Coupled biogeochemical and hydrological responses of streams and rivers to drought. *Freshwater Biology* 48:1219-1231.

Dangles, O.; Gessner, M. O.; Guerold, F. & Chauvet, E. **2004** Impacts of stream acidification on litter breakdown: implications for assessing ecosystem functioning. *Journal of Applied Ecology* 41:365-378.

Das, M.; Royer, T. V. & Leff, L. G. **2012** Interactions between aquatic bacteria and an aquatic hyphomycete on decomposing maple leaves. *Fungal Ecology* 5:236-244.

Datry, T.; Corti, R.; Claret, C. & Philippe, M. **2011** Flow intermittence controls leaf litter breakdown in a French temporary alluvial river: the "drying memory". *Aquatic Sciences* 73:471-483.

Day, T. A.; Zhang, E. T. & Ruhland, C. T. **2007** Exposure to solar UV-B radiation accelerates mass and lignin loss of *Larrea tridentata* litter in the Sonoran Desert. *Plant Ecology* 193:185-194.

de Groot, C.-J. & Fabre, A. **1993** The impact of desiccation of a freshwater marsh (Garcines Nord, Camargue, France) on sediment-water-vegetation interactions. Part 3: The fractional composition and the phosphate adsorption characteristics of the sediment. *Hydrobiologia* 252:105-116.

de Groot, C.-J. & van Wijck, C. **1993** The impact of desiccation of a freshwater marsh (Garcines Nord, Camargue, France) on sediment-water-vegetation interactions Part 1: The sediment chemistry. *Hydrobiologia* 252:83-94.

de Vicente, I.; Andersen, F. O.; Hansen, H. C. B.; Cruz-Pizarro, L. & Jensen, H. S. **2010** Water level fluctuations may decrease phosphate adsorption capacity of the sediment in oligotrophic high mountain lakes. *Hydrobiologia* 651:253-264.

de Vicente, I.; López, R.; Pozo, I. & Green, A. J. **2012** Nutrient and sediment dynamics in a Mediterranean shallow lake in southwest Spain. *Limnetica* 31:231-250.

Denward, C. & Tranvik, L. **1998** Effects of solar radiation on aquatic macrophyte litter decomposition. *Oikos* 82:51-58.

Denward, C.; Anesio, A.; Graneli, W. & Tranvik, L. **2001** Solar radiation effects on decomposition of macrophyte litter in a lake littoral. *Archiv für Hydrobiologie* 152:69-80.

Dieter, D.; von Schiller, D.; García-Roger, E. M.; Sánchez-Montoya, M. M.; Gómez, R.; Mora-Gómez, J.; Sangiorgio, F.; Gelbrecht, J. & Tockner, K. **2011** Preconditioning effects of intermittent stream flow on leaf litter decomposition. *Aquatic Sciences* 73:599-609.

Dittrich, M.; Chesnyuk, A.; Gudimov, A.; McCulloch, J.; Quazi, S.; Young, J.; Winter, J.; Stainsby, E. & Arhonditsis, G. **2013** Phosphorus retention in a mesotrophic lake under transient loading conditions: Insights from a sediment phosphorus binding form study. *Water Research* 47:1433-1447.

Driescher, E.; Behrendt, H.; Schellenberger, G. & Stellmacher, R. **1993** Lake Müggelsee and its environment - Natural conditions and anthropogenic impacts. *Internationale Revue Der Gesamten Hydrobiologie* 78:327-343.

References

Dudgeon, D. & Wu, K. K. Y. **1999** Leaf litter in a tropical stream: food or substrate for macroinvertebrates? *Archiv für Hydrobiologie* 146:65-82.

Einsele, W. **1936** Über die Beziehungen des Eisenkreislaufs zum Phosphatkreislauf im eutrophen See. *Archiv für Hydrobiologie* 29:664–686.

Einsele, W. **1938** Über chemische und kolloidchemische Vorgänge in Eisen-Phosphat- Systemen unter limnischen and limnogeologischen Gesichtpunkten. *Archiv für Hydrobiologie* 33:361–387.

Ellis, L. M.; Molles, M. C. & Crawford, C. S. **1999** Influence of experimental flooding on litter dynamics in a Rio Grande riparian forest, New Mexico. *Restoration Ecology* 7:193-204.

European Commission **2012** Report to the European Parliament and the Council on the implementation of the Water Framework Directive - River Basin Management Plans. *Brussels, Belgium*.

European Parliament and the Council **2000** Directive 2000/60/EC on establishing a framework for community action in the field of water policy. *Brussels, Belgium*.

European Parliament and the Council **2007** Directive 2007/60/EC on the assessment and management of flood risks. *Brussels, Belgium*.

Fabre, A. **1988** Experimental studies on some factors influencing phosphorus solubilization in connection with the drawdown of a reservoir. *Hydrobiologia* 159:153-158.

Fellman, J. B.; Petrone, K. C. & Grierson, P. F. **2013** Leaf litter age, chemical quality, and photodegradation control the fate of leachate dissolved organic matter in a dryland river. *Journal of Arid Environments* 89:30-37.

Fenoglio, S.; Bo, T.; Cucco, M. & Malacarne, G. **2006** Leaf breakdown patterns in a NW Italian stream: Effect of leaf type, environmental conditions and patch size. *Biologia* 61:555-563.

References

Fernández-Aláez, M. & Fernández-Aláez, C. **2010** Effects of the intense summer desiccation and the autumn filling on the water chemistry in some Mediterranean ponds. *Limnetica* 29:59-73.

Ferreira, V. & Chauvet, E. **2012** Changes in dominance among species in aquatic hyphomycete assemblages do not affect litter decomposition rates. *Aquatic Microbial Ecology* 66:1-11.

Fischer, H.; Mille-Lindblom, C.; Zwirnmann, E. & Tranvik, L. J. **2006** Contribution of fungi and bacteria to the formation of dissolved organic carbon from decaying common reed (*Phragmites australis*). *Archiv für Hydrobiologie* 166:79-97.

Fox, L. E. **1989** A model for inorganic control of phosphate concentrations in river waters. *Geochimica et Cosmochimica Acta* 53:417-428.

Fromin, N.; Pinay, G.; Montuelle, B.; Landais, D.; Ourcival, J. M.; Joffre, R. & Lensi, R. **2010** Impact of seasonal sediment desiccation and rewetting on microbial processes involved in greenhouse gas emissions. *Ecohydrology* 3:339-348.

Gächter, R. **1987** Lake restoration - why oxygenation and artificial mixing cannot substitute for a decrease in the external phosphorus loading. *Schweizerische Zeitschrift für Hydrologie - Swiss Journal Of Hydrology* 49:170-185.

Gallart, F.; Prat, N.; García-Roger, E. M.; Latron, J.; Rieradevall, M.; Llorens, P.; Barbera, G. G.; Brito, D.; de Girolamo, A. M.; Lo Porto; et al. **2012** A novel approach to analysing the regimes of temporary streams in relation to their controls on the composition and structure of aquatic biota. *Hydrology and Earth System Sciences* 16:3165-3182.

García, C. M. & Niell, F. X. **1993** Seasonal change in a saline temporary lake (Fuente de Piedra, southern Spain). *Hydrobiologia* 267:211-223.

References

Gelbrecht, J.; Lengsfeld, H.; Pöthig, R. & Opitz, D. **2005** Temporal and spatial variation of phosphorus input, retention and loss in a small catchment of NE Germany. *Journal of Hydrology* 304:151–165.

Geraldes, A. M. & Boavida, M. J. L. **2004** Limnological variations of a reservoir during two successive years: One wet, another dry. *Lakes and Reservoirs: Research and Management* 9:143-152.

Germer, S.; Kaiser, K.; Bens, O. & Hüttl, R. F. **2011** Water balance changes and responses of ecosystems and society in the berlin-brandenburg region - A review. *Die Erde* 142:65-95.

Gessner, M. O. & Chauvet, E. **1993** Ergosterol-to-biomass conversion factors for aquatic hyphomycetes. *Applied and Environmental Microbiology* 59:502-507.

Gessner, M. O. & Chauvet, E. **1994** Importance of stream microfungi in controlling breakdown rates of leaf-litter. *Ecology* 75:1807-1817.

Gessner, M. O. & Chauvet, E. **2002** A case for using litter breakdown to assess functional stream integrity. *Ecological Applications* 12:498-510.

Gessner, M. O. **1991** Differences in processing dynamics of fresh and dried leaf litter in a stream ecosystem. *Freshwater Biology* 26:387-398.

Gessner, M. O. **2005a** Ergosterol as a measure of fungal biomass. In: Graça, M. A. S.; Bärlocher, F. & Gessner, M. O. (eds.) Methods to study litter decomposition: A practical guide. *Springer, Dordrecht, Netherlands,* 189-195.

Gessner, M. O. **2005b** Proximate lignin and cellulose. In: Graça, M.; Bärlocher, F. & Gessner, M. O. (eds.) Methods to study litter decomposition: A practical guide. *Springer, Dordrecht, Netherlands,* 115-120.

Gessner, M. O.; Chauvet, E. & Dobson, M. **1999** A perspective on leaf litter breakdown in streams. *Oikos* 85:377-384.

References

Giorgi, F.; Bi, X. Q. & Pal, J. **2004** Mean, interannual variability and trends in a regional climate change experiment over Europe. II: climate change scenarios (2071-2100). *Climate Dynamics* 23:839-858.

Glazebrook, H. S. & Robertson, A. I. **1999** The effect of flooding and flood timing on leaf litter breakdown rates and nutrient dynamics in a river red gum (*Eucalyptus camaldulensis*) forest. *Australian Journal of Ecology* 24:625-635.

Graça, M. A. S. **2001** The role of invertebrates on leaf litter decomposition in streams - A review. *International Review of Hydrobiology* 86:383-393.

Grierson, P. F.; Comerford, N. B. & Jokela, E. J. **1998** Phosphorus mineralization kinetics and response of microbial phosphorus to drying and rewetting in a Florida Spodosol. *Soil Biology & Biochemistry* 30:1323-1331.

Gulis, V. & Suberkropp, K. F. **2003** Leaf litter decomposition and microbial activity in nutrient-enriched and unaltered reaches of a headwater stream. *Freshwater Biology* 48:123-134.

Gulis, V. & Suberkropp, K. F. **2006** Fungi: biomass, production, and sporulation of aquatic hyphomycetes. In: Hauer, F. R. & Lamberti, G. A. (eds.) Methods in stream ecology. *Elsevier Academic Press, Burlington, USA*, 311-325.

Güsewell, S. & Gessner, M. O. **2009** N:P ratios influence litter decomposition and colonization by fungi and bacteria in microcosms. *Functional Ecology* 23:211-219.

Haase, K. & Wantzen, K. M. **2008** Analysis and decomposition of condensed tannins in tree leaves. *Environmental Chemistry Letters* 6:71-75.

Hagerman, A. E. **2011** The tannin handbook. http://www.users.muohio.edu/hagermae/ (last verified 12 April 2013)

References

Henry, H. A. L.; Brizgys, K. & Field, C. B. **2008** Litter decomposition in a Californian annual grassland: Interactions between photodegradation and litter layer thickness. *Ecosystems* 11:545-554.

Hladyz, S.; Gessner, M. O.; Giller, P. S.; Pozo, J. & Woodward, G. **2009** Resource quality and stoichiometric constraints on stream ecosystem functioning. *Freshwater Biology* 54:957-970.

Holmer, M. & Storkholm, P. **2001** Sulphate reduction and sulphur cycling in lake sediments: a review. *freshwater Biology* 46:431-451.

Hulthe, G.; Hulth, S. & Hall, P. O. J. **1998** Effect of oxygen on degradation rate of refractory and labile organic matter in continental margin sediments. *Geochimica et Cosmochimica Acta* 62:1319-1328.

Humphries, P. & Baldwin, D. S. **2003** Drought and aquatic ecosystems: an introduction. *Freshwater Biology* 48:1141-1146.

Hupfer, M. & Gächter, R. **1995** Polyphosphate in lake-sediments - P-31 NMR-spectroscopy as a tool for its identification. *Limnology And Oceanography* 40:610-617.

Hupfer, M. & Lewandowski, J. **2008** Oxygen controls the phosphorus release from lake sediments - a long-lasting paradigm in Limnology. *International Review of Hydrobiology* 93:415-432.

Hupfer, M. **1995** Bindungsformen und Mobilität des Phosphors in Gewässersedimenten (IV-3.2). In: Steinberg, C.; Bernhardt, H. & Klapper, H. (eds.) Handbuch Angewandte Limnologie. *ecomed-Verlag, Landsberg am Lech, Germany*, 1-22.

Hupfer, M.; Gächter, R. & Giovanoli, R. **1995** Transformation of phosphorus species in settling seston and during early sediment diagenesis. *Aquatic Sciences* 57: 305-324.

References

Hupfer, M.; Gloess, S. & Grossart, H.-P. **2007** Polyphosphate-accumulating microorganisms in aquatic sediments. *Aquatic Microbial Ecology* 47:299–311.

Ihaka, R. & Gentleman, R. **1996** R: a language for data analysis and graphics. *Journal of Computational and Graphical Statistics* 5:299-314.

Irons, J. G.; Oswood, M. W.; Stout, R. J. & Pringle, C. M. **1994** Latitudinal patterns in leaf-litter breakdown - is temperature really important. *Freshwater Biology* 32:401-411.

Jensen, H. S. & Andersen, F. O. **1992** Importance of temperature, nitrate, and pH for phosphate release from aerobic sediments of 4 shallow, eutrophic lakes. *Limnology and Oceanography* 37:577-589.

Jensen, H. S.; Kristensen, P.; Jeppesen, E. & Skytthe, A. **1992** Iron-phosphorus ratio in surface sediment as an indicator of phosphate release from aerobic sediments in shallow lakes. *Hydrobiologia* 235:731-743.

Junk, W. J.; Bayley, P. B. & Sparks, R. E. **1989** The flood pulse concept in river-floodplain system. *Proceedings of the International Large Rivers Symposium. Canadian special publication of Fisheries and Aquatic Sciences* 106:110-127.

Kamp-Nielsen, L. **1974** Mud-water exchange of phosphate and other ions in undisturbed sediment cores and factors affecting the exchange rates. *Archiv für Hydrobiologie* 73:218-237.

Kerr, J. G.; Burford, M.; Olley, J. & Udy, J. **2010** The effects of drying on phosphorus sorption and speciation in subtropical river sediments. *Marine and Freshwater Research* 61:928-935.

Kleeberg, A. & Kohl, J.-G. **1999** Assessment of the long-term effectiveness of sediment dredging to reduce benthic phosphorus release in shallow Lake Muggelsee (Germany). *Hydrobiologia* 394:153-161.

References

Kleeberg, A. & Kozerski, H. **1997** Phosphorus release in lake Grosser Müggelsee and its implications for lake restoration. *Hydrobiologia* 342:9-26.

Kleeberg, A. **2002** Phosphorus sedimentation in seasonal anoxic Lake Scharmützel, NE Germany. *Hydrobiologia* 472:53-65.

Kleeberg, A.; Herzog, C.; Jordan, S. & Hupfer, M. **2010** What drives the evolution of the sedimentary phosphorus cycle? *Limnologica* 40:102-113.

Kowalchuk, G. A.; Stephen, J.; DeBoer, W.; Prosser, J. I.; Embley, T. M. & Woldendorp, J. W. **1997** Analysis of ammonia-oxidizing bacteria of the beta subdivision of the class *Proteobacteria* in coastal sand dunes by denaturing gradient gel electrophoresis and sequencing of PCR-amplified 16S ribosomal DNA fragments. *Applied and Environmental Microbiology* 63:1489-1497.

Kraus, T. E. C.; Dahlgren, R. A. & Zasoski, R. J. **2003** Tannins in nutrient dynamics of forest ecosystems - a review. *Plant and Soil* 256:41-66.

Küsel, K. & Drake, H. L. **1996** Anaerobic capacities of leaf litter. *Applied and Environmental Microbiology* 62:4216-4219.

Lake, P. S. **2003** Ecological effects of perturbation by drought in flowing waters. *Freshwater Biology* 48:1161-1172.

Lake, P. S. **2011** Drought and Aquatic Ecosystems - Effects and Responses. *Wiley-Blackwell, Wiley & Sons Ltd., West Sussex, UK,* 400p.

Langhans, S. D. & Tockner, K. **2006** The role of timing, duration, and frequency of inundation in controlling leaf litter decomposition in a river-floodplain ecosystem (Tagliamento, northeastern Italy). *Oecologia* 147:501-509.

Langhans, S. D.; Tiegs, S. D.; Gessner, M. O. & Tockner, K. **2008** Leaf-decomposition heterogeneity across a riverine floodplain mosaic. *Aquatic Science* 70:337-346.

Larned, S. T.; Datry, T.; Arscott, D. B. & Tockner, K. **2010** Emerging concepts in temporary river ecology. *Freshwater Biology* 55:717–738.

References

Laskov, C.; Herzog, C.; Lewandowski, J. & Hupfer, M. **2007** Miniaturized photometrical methods for the rapid analysis of phosphate, ammonium, ferrous iron, and sulfate in pore water of freshwater sediments. *Limnology and Oceanography: Methods* 4:63–71.

Lee, S. H.; Malone, C. & Kemp, P. F. **1993** Use of multiple 16S ribosomal-RNA-targeted fluorescent-probes to increase signal strength and measure cellular RNA from natural planktonic bacteria. *Marine Ecology Progress Series* 101:193-201.

Leopold, A. C.; Musgrave, M. E. & Williams, K. M. **1981** Solute leakage resulting from leaf desiccation. *Plant Physiology* 68:1222-1225.

Leroy, C. J. & Marks, J. C. **2006** Litter quality, stream characteristics and litter diversity influence decomposition rates and macroinvertebrates. *Freshwater Biology* 51:605-617.

Lewandowski, J. & Hupfer, M. **2005** Effect of macrozoobenthos on two-dimensional small-scale heterogeneity of pore water phosphorus concentrations in lake sediments: A laboratory study. *Limnology and Oceanography* 50:1106-1118.

Li, A. O. Y.; Ng, L. C. Y. & Dudgeon, D. **2009** Effects of leaf toughness and nitrogen content on litter breakdown and macroinvertebrates in a tropical stream. *Aquatic Sciences* 71:80-93.

Libkind, D.; Brizzio, S.; Ruffini, A.; Gadanho, M.; van Broock, M. & Sampaio, J. **2003** Molecular characterization of carotenogenic yeasts from aquatic environments in Patagonia, Argentina. *Antonie van Leeuwenhoek* 84:313-322.

Lijklema, L. **1980** Interaction of orthophosphate with iron(III) and aluminum hydroxides. *Environmental Science and Technology* 14:537–541.

References

Liu, Y.; Villalba, G.; Ayres, R. U. & Schroder, H. **2008** Global phosphorus flows and environmental impacts from a consumption perspective. *Journal of Industrial Ecology* 12:229-247.

Loeb, R.; Lamers, L. P. M. & Roelofs, J. G. M. **2008** Effects of winter versus summer flooding and subsequent desiccation on soil chemistry in a riverine hay meadow. *Geoderma* 145:84-90.

Loranger, G.; Ponge, J. F.; Imbert, D. & Lavelle, P. **2002** Leaf decomposition in two semi-evergreen tropical forests: influence of litter quality. *Biology and Fertility of Soils* 35:247-252.

Maamri, A.; Bärlocher, F.; Pattee, E. & Chergui, H. **2001** Fungal and bacterial colonisation of *Salix pedicellata* leaves decaying in permanent and intermittent streams in eastern Morocco. *International Reviews in Hydrobiology* 86:337–348.

Maamri, A.; Chergui, H. & Pattee, E. **1997** Leaf litter processing in a temporary northeastern Moroccan river. *Archiv für Hydrobiologie* 140:513-531.

Mainstone, C. P. & Parr, W. **2002** Phosphorus in rivers - ecology and management. *Science of the Total Environment* 282:25-47.

Mariotti, A.; Zeng, N.; Yoon, J.-H.; Artale, V.; Navarra, A.; Alpert, P. & Li, L. Z. X. **2008** Mediterranean water cycle changes: transition to drier 21st century conditions in observations and CMIP3 simulations. *Environmental Research Letters* 3, doi:10.1088/1748-9326/3/4/044001.

McClain, M. E.; Boyer, E. W.; Dent, C. L.; Gergel, S. E.; Grimm, N. B.; Groffman, P. M.; Hart, S. C.; Harvey, J. W.; Johnston, C. A.; Mayorga, E. et al. **2003** Biogeochemical hot spots and hot moments at the interface of terrestrial and aquatic ecosystems. *Ecosystems* 6:301-312.

McComb, A. J.; Qiu, S.; Bell, R. W. & Davis, J. A. **2007** Catchment litter: a phosphorus source mobilized during seasonal rainfall. *Nutrient Cycling in Agroecosystems* 77:179-186.

References

McIntyre, R. E. S.; Adams, M. A.; Ford, D. J. & Grierson, P. F. **2009** Rewetting and litter addition influence mineralisation and microbial communities in soils from a semi-arid intermittent stream. *Soil Biology and Biochemistry* 41:92-101.

Meehl, G. A.; Stocker, T. F.; Collins, W. D.; Friedlingstein, P.; Gaye, A. T.; Gregory, J. M.; Kitoh, A.; Knutti, R.; Murphy, J. M.; Noda, A.; et al. **2007** Global Climate Projections. In: Solomon, S.; Qin, D.; Manning, M.; Chen, Z.; Marquis, M.; Averyt, K. B.; Tignor, M. & Miller, H. (eds.) Climate Change 2007: The Physical Science Basis. Contribution of Working Group I to the Fourth Assessment Report of the Intergovernmental Panel on Climate Change. *Cambridge University Press, Cambridge, UK and New York, USA*, 747-846.

Meentemeyer, V. **1978** Approach to biometeorology of decomposer organisms. *International Journal of Biometeorology* 22:94-102.

Mitchell, A. & Baldwin, D. S. **1998** Effects of desiccation/oxidation on the potential for bacterially mediated P release from sediments. *Limnology and oceanography* 43:481-487.

Moore, P. & Reddy, K. **1994** Role of Eh and pH on phosphorus geochemistry in sediments of lake Okeechobee, Florida. *Journal of Environmental Quality* 23:955-964.

Mortimer, C. H. **1941, 1942** The exchange of dissolved substances between mud and water in lakes. *Journal of Ecology* 29, 30:280-329, 147-201.

Murphy, J. & Riley, J. P. **1962** A modified single solution method for determination of phosphate in natural waters. *Analytica Chimica Acta* 27:31-36.

Muyzer, G.; Dewaal, E. C. & Uitterlinden, A. G. **1993** Profiling of complex microbial populations by denaturing gradient gel-electrophoresis analysis

References

of polymerase chain reaction-amplified genes-coding for 16S ribosomal-RNA. *Applied and Environmental Microbiology* 59:695-700.

Nair, P. S.; Logan, T. J.; Sharpley, A. N.; Sommers, L. E.; Tabatabai, M. A. & Yuan, T. L. **1984** Interlaboratory comparison of a standardized phosphorus adsorption procedure. *Journal of Enviromental Quality* 13:591-595.

Naselli-Flores, L. & Barone, R. **2005** Water-level fluctuations in Mediterranean reservoirs: Setting a dewatering threshold as a management tool to improve water quality. *Hydrobiologia* 548:85-99.

Nercessian, O.; Noyes, E.; Kalyuzhnaya, M. G.; Lidstrom, M. E. & Chistoserdova, L. **2005** Bacterial populations active in metabolism of C-1 compounds in the sediment of Lake Washington, a freshwater lake. *Applied and Environmental Microbiology* 71:6885-6899.

Newbold, J. D.; Elwood, J. W.; Oneill, R. V. & Sheldon, A. L. **1983** Phosphorus dynamics in a woodland stream ecosystem - a study of nutrient spiralling. *Ecology* 64:1249-1265.

Newbold, J. D.; Elwood, J. W.; Oneill, R. V. & Vanwinkle, W. **1981** Measuring nutrient spiralling in streams. *Canadian Journal of Fisheries and Aquatic Sciences* 38:860-863.

Nguyen, B. T. & Marschner, P. **2005** Effect of drying and rewetting on phosphorus transformations in red brown soils with different soil organic matter content. *Soil Biology & Biochemistry* 37:1573-1576.

Nosanchuk, J. D. & Casadevall, A. **2003** The contribution of melanin to microbial pathogenesis. *Cell Microbiology* 5:203-223.

Nowlin, W. H.; Davies, J.-M.; Nordin, R. N. & Mazumder, A. **2004** Effects of water level fluctuation and short-term climate variation on thermal and stratification regimes of a British Columbia reservoir and lake. *Lake and Reservoir Management* 20:91-109.

References

Obermann, M.; Froebrich, J.; Perrin, J.-L. & Tournoud, M.-G. **2007** Impact of significant floods on the annual load in an agricultural catchment in the Mediterranean. *Journal of Hydrology* 334:99–108.

Oksanen, J.; Blanchet, F. G.; Kindt, R.; Legendre, P.; O'Hara, R. B.; Simpson, G. L.; Solymos, P.; Stevens, M. H. H. & Wagner, H. **2011** vegan: community ecology package. R package version 1.17-10. http://CRAN.R-project.org/package=vegan

Ostendorp, W. & Frevert, T. **1979** Untersuchungen zur Manganfreisetzung und zum Mangangehalt der Sedimentoberschicht im Bodensee. *Archiv für Hydrobiologie* 55:255-277.

Ostrofsky, M. **1997** Relationship between chemical characteristics of autumn-shed leaves and aquatic processing rates. *Journal of the North American Benthological Society* 16:750-759.

Pabst, S.; Scheifhacken, N.; Hesselschwerdt, J. & Wantzen, K. M. **2008** Leaf litter degradation in the wave impact zone of a pre-alpine lake. *Hydrobiologia* 613:117-131.

Petersen, R. C. & Cummins, K. W. **1974** Leaf processing in a woodland stream. *Freshwater Biology* 4:343-368.

Pettersson, K.; Boström, B. & Jacobsen, O. S. **1988** Phosphorus in sediments - Speciation and analysis. *Hydrobiologia* 170:91-101.

Prat, N.; Gallart, F.; García-Roger, E. M.; Latron, J.; Rieradevall, M.; Llorens, P.; Barberá, G. G.; Brito, D.; de Girolamo, A. M.; Dieter D.; et al. **2013** The Mirage Toolbox: An integrated assessment tool for temporary streams. *River Research and Applications (Temporary Streams)*, submitted.

Psenner, R.; Pucsko, R. & Sager, M. **1984** Fractionation of organic and inorganic phosphorous compounds in lake sediments - An attempt to characterize ecologically important fractions. *Archiv für Hydrobiologie* 70:111-155.

References

Qiu, S. & McComb, A. J. **1994** Effects of oxygen concentration on phosphorus release from reflooded air-dried wetland sediments. *Australian Journal of Marine and Freshwater Research* 45:1319-1328.

Qiu, S. & McComb, A. J. **2002** Interrelations between iron extractability and phosphate sorption in reflooded air-dried sediments. *Hydrobiologia* 472:39-44.

Qiu, S.; McComb, A. J.; Bell, R. W. & Davis, J. A. **2004** Phosphorus dynamics from vegetated catchment to lakebed during seasonal refilling. *Wetlands* 24:828-836.

Raich, J. W. & Schlesinger, W. H. **1992** The global carbon-dioxide flux in soil respiration and its relationship to vegetation and climate. *Tellus Series B - Chemical and Physical Meteorology* 44:81-99.

Raskin, L.; Poulsen, L. K.; Noguera, D. R.; Rittmann, B. E. & Stahl, D. A. **1994** Quantification of methanogenic groups in anaerobic biological reactors by oligonucleotide probe hybridization. *Applied and Environmental Microbiology* 60:1241-1248.

Raviraja, N.; Nikolcheva, L. & Bärlocher, F. **2005** Diversity of conidia of aquatic hyphomycetes assessed by microscopy and by DGGE. *Microbial Ecology* 49:301-307.

Reddy, K. R.; Kadlec, R. H.; Flaig, E. & Gale, P. M. **1999** Phosphorus retention in streams and wetlands: A review. *Critical Reviews in Environmental Science and Technology* 29:83-146.

Reith, F.; Drake, H. L. & Küsel, K. **2002** Anaerobic activities of bacteria and fungi in moderately acidic conifer and deciduous leaf litter. *FEMS Microbiology Ecology* 41:27-35.

Richardson, J. S. **1992** Food, microhabitat, or both - macroinvertebrate use of leaf accumulations in a montane stream. *Freshwater Biology* 27:169-176.

References

Roden, E. E. & Edmonds, J. W. **1997** Phosphate mobilization in iron-rich anaerobic sediments: Microbial Fe(III) oxide reduction versus iron-sulfide formation. *Archiv für Hydrobiologie* 139:347-378.

Romaní, A. M.; Fischer, H.; Mille-Lindblom, C. & Tranvik, L. J. **2006** Interactions of bacteria and fungi on decomposing litter: Differential extracellular enzyme activities. *Ecology* 87:2559-2569.

Royer, T. V. & Minshall, G. W. **2001** Effects of nutrient enrichment and leaf quality on the breakdown of leaves in a hardwater stream. *Freshwater Biology* 46:603-610.

Salmanowicz, B. & Nylund, J.-E. **1988** High-performance liquid-chromatography determination of ergosterol as a measure of ectomycorrhiza infection in scots pine. *European Journal of Forest Pathology* 18:291-298.

Sariyildiz, T. & Anderson, J. M. **2003** Interactions between litter quality, decomposition and soil fertility: a laboratory study. *Soil Biology and Biochemitstry* 35:391-399.

Schlief, J. & Mutz, M. **2007** Response of aquatic leaf associated microbial communities to elevated leachate DOC: A microcosm study. *International Review of Hydrobiology* 92:146-155.

Schlief, J. & Mutz, M. **2011** Leaf decay processes during and after a supra-seasonal hydrological drought in a temperate lowland stream. *International Review of Hydrobiology* 96:633-655.

Schmidt-Kloiber, A.; Graf, W.; Lorenz, A. & Moog, O. **2006** The AQEM/STAR taxalist - a pan-European macro-invertebrate ecological database and taxa inventory. *Hydrobiologia* 566:325-342.

Schönbrunner, I. M.; Preiner, S. & Hein, T. **2012** Impact of drying and re-flooding of sediment on phosphorus dynamics of river-floodplain systems. *Science of the Total Environment* 432:329-337.

References

Schwoerbel, J. **1999** Einführung in die Limnologie. *8th edn. Gustav Fischer Verlag, Stuttgart, Germany*, 464pp.

Song, K.-Y.; Zoh, K.-D. & Kang, H. **2007** Release of phosphate in a wetland by changes in hydrological regime. *Science of the Total Environment* 380:13-18.

Sperber, J. I. **1958** Release of phosphate from soil minerals by hydrogen sulphide. *Nature* 181:934.

Sridhar, K. R.; Duarte, S.; Cassio, F. & Pascoal, C. **2009** The role of early fungal colonizers in leaf-litter decomposition in portuguese streams impacted by agricultural runoff. *International Review of Hydrobiology* 94:399-409.

Stanley, E. H.; Fisher, S. G. & Grimm, N. B. **1997** Ecosystem expansion and contraction in streams. *Bioscience* 47:427-435.

Steger, K.; Premke, K.; Gudasz, C.; Sundh, I. & Tranvik, L. J. **2011** Microbial biomass and community composition in boreal lake sediments. *Limnology and Oceanography* 56:725-733.

Steward, A. L.; Marshall, J. C.; Sheldon, F.; Harch, B.; Choy, S.; Bunn, S. E. & Tockner, K. **2011** Terrestrial invertebrates of dry river beds are not simply subsets of riparian assemblages. *Aquatic Sciences* 73:551-566.

Steward, A. L.; von Schiller, D.; Tockner, K.; Marshall, J. C. & Bunn, S. E. **2012** When the river runs dry: human and ecological values of dry riverbeds. *Frontiers in Ecology and the Environment* 10:202-209.

Taylor, B. R. & Bärlocher, F. **1996** Variable effects of air-drying on leaching losses from tree leaf litter. *Hydrobiologia* 325:173-182.

Teske, A.; Sigalevich, P.; Cohen, Y. & Muyzer, G. **1996** Molecular identification of bacteria from a coculture by denaturing gradient gel electrophoresis of 16S ribosomal DNA fragments as a tool for isolation in pure cultures. *Applied and Environmental Microbiology* 62:4210-4215.

References

Throback, I. N.; Enwall, K.; Jarvis, A. & Hallin, S. **2004** Reassessing PCR primers targeting nirS, nirK and nosZ genes for community surveys of denitrifying bacteria with DGGE. *FEMS Microbiology Ecology* 49:401-417.

Tockner, K.; Uehlinger, U.; Robinson, C. T.; Siber, R.; Tonolla, D. & Peter, F. D. **2009** European Rivers. In: Likens, G. E. (ed.) Encyclopedia of Inland Waters. *1st edn. Elsevier Academic Press, Waltham, USA*, 366-377.

Tuomi, M.; Thum, T.; Arvinen, H. J.; Fronzek, S.; Berg, B.; Harmon, M.; Trofymow, J. A.; Sevanto, S. & Liski, J. **2009** Leaf litter decomposition – estimates of global variability based on yasso07 model. *Ecological Modelling* 220:3362-3371.

Turner, B. L. & Haygarth, P. M. **2001** Biogeochemistry - Phosphorus solubilization in rewetted soils. *Nature* 411:258.

Turner, B. L.; Driessen, J. P.; Haygarth, P. M. & Mckelvie, I. D. **2003** Potential contribution of lysed bacterial cells to phosphorus solubilisation in two rewetted Australian pasture soils. *Soil Biology & Biochemistry* 35:187-189.

Twinch, A. J. **1987** Phosphate exchange characteristics of wet and dried sediment samples from a hypertrophic reservoir: Implications for the measurements of sediment phosphorus status. *Water Research* 21:1225-1230.

Väisänen, A.; Laatikainen, P.; Ilander, A. & Renvall, S. **2008** Determination of mineral and trace element concentrations in pine needles by ICP-OES: evaluation of different sample pre-treatment methods. *International Journal of Environmental Analytical Chemistry* 88:1005-1016.

Vannote, R. L.; Minshall, G. W.; Cummins, K. W.; Sedell, J. R. & Cushing, C. E. **1980** The river continuum concept. *Canadian Journal of Fisheries and Aquatic Sciences* 37:130-137.

References

von Schiller, D.; Acuña, V.; Graeber, D.; Marti, E.; Ribot, M.; Sabater, S.; Timoner, X. & Tockner, K. **2011** Contraction, fragmentation and expansion dynamics determine nutrient availability in a Mediterranean forest stream. *Aquatic Sciences* 73:485-497.

von Schiller, D.; Martí, E.; Riera, J. L.; Ribot, M.; Argerich, A.; Fonollà, P. & Sabater, F. **2008** Inter-annual, annual, and seasonal variation of P and N retention in a perennial and an intermittent stream. *Ecosystems* 11:670-687.

Wantzen, K. M.; Junk, W. J. & Rothhaupt, K.-O. **2008a** An extension of the floodpulse concept (FPC) for lakes. *Hydrobiologia* 613:151-170.

Wantzen, K. M.; Rothhaupt, K.-O.; Moertl, M.; Cantonati, M.; G.-Toth, L. & Fischer, P. **2008b** Ecological effects of water-level fluctuations in lakes: an urgent issue. *Hydrobiologia* 613:1-4.

Watts, C. J. **2000a** Seasonal phosphorus release from exposed, re-inundated littoral sediments of two Australian reservoirs. *Hydrobiologia* 431:27-39.

Watts, C. J. **2000b** The effect of organic matter on sedimentary phosphorus release in an Australian reservoir. *Hydrobiologia* 431:13-25.

Webster, J. R. & Benfield, E. F. **1986** Vascular plant breakdown in fresh-water ecosystems. *Annual Review of Ecology and Systematics* 17:567-594.

Williams, D. D. **2006** The biology of temporary waters. *Oxford University Press, New York*, 348pp.

Wilson, J. S. & Baldwin, D. S. **2008** Exploring the "Birch effect" in reservoir sediments: influence of inundation history on aerobic nutrient release. *Chemistry and Ecology* 24:379-386.

Wishart, M. J. **2000** The terrestrial invertebrate fauna of a temporary stream in southern Africa. *African Zoology* 35:193-200.

Withers, P. J. A. & Jarvie, H. P. **2008** Delivery and cycling of phosphorus in rivers: A review. *Science of the Total Environment* 400:379-395.

References

Woodward, G.; Gessner, M. O.; Giller, P. S.; Gulis, V.; Hladyz, S.; Lecerf, A.; Malmqvist, B.; McKie, B. G.; Tiegs, S. D.; Cariss, H.; et al. **2012** Continental-scale effects of nutrient pollution on stream ecosystem functioning. *Science* 336:1438-1440.

Worsfold, P. J.; Monbet, P.; Tappin, A. D.; Fitzsimons, M. F.; Stiles, D. A. & McKelvie, I. D. **2008** Characterisation and quantification of organic phosphorus and organic nitrogen components in aquatic systems: A review. *Analytica Chimica Acta* 624:37-58.

Xiao, W.-J.; Song, C.-L.; Cao, X.-Y. & Zhou, Y.-Y. **2012** Effects of air-drying on phosphorus sorption in shallow lake sediment, China. *Fresenius Environmental Bulletin* 21:672-678.

Yoshimura, C.; Gessner, M. O.; Tockner, K. & Furumai, H. **2008** Chemical properties, microbial respiration, and decomposition of coarse and fine particulate organic matter. *Journal of The North American Benthological Society* 27:664-673.

Young, J. C. **1995** Microwave-assisted extraction of the fungal metabolite ergosterol and total fatty-acids. *Journal of Agricultural and Food Chemistry* 43:2904-2910.

Young, R. G.; Matthaei, C. D. & Townsend, C. R. **2008** Organic matter breakdown and ecosystem metabolism: functional indicators for assessing river ecosystem health. *Journal of The North American Benthological Society* 27:605-625.

Zak, D. & Gelbrecht, J. **2007** The mobilisation of phosphorus, organic carbon and ammonium in the initial stage of fen rewetting (a case study from NE Germany). *Biogeochemistry* 85:141–151.

Zak, D.; Kleeberg, A. & Hupfer, M. **2006** Sulphate-mediated phosphorus mobilization in riverine sediments at increasing sulphate concentration, River Spree, NE Germany. *Biogeochemistry* 80:109-119.

References

Zak, D.; Wagner, C.; Payer, B.; Augustin, J. & Gelbrecht, J. **2010** Phosphorus mobilization in rewetted fens: the effect of altered peat properties and implications for their restoration. *Ecological Applications* 20:1336-1349.

Zhang, D.; Hui, D.; Luo, Y. & Zhou, G. **2008** Rates of litter decomposition in terrestrial ecosystems: global patterns and controlling factors. *Journal of Plant Ecology* 1:85-93.

i want morebooks!

Buy your books fast and straightforward online - at one of world's fastest growing online book stores! Environmentally sound due to Print-on-Demand technologies.

Buy your books online at
www.get-morebooks.com

Kaufen Sie Ihre Bücher schnell und unkompliziert online – auf einer der am schnellsten wachsenden Buchhandelsplattformen weltweit! Dank Print-On-Demand umwelt- und ressourcenschonend produziert.

Bücher schneller online kaufen
www.morebooks.de

 VDM Verlagsservicegesellschaft mbH
Heinrich-Böcking-Str. 6-8   Telefon: +49 681 3720 174   info@vdm-vsg.de
D - 66121 Saarbrücken   Telefax: +49 681 3720 1749   www.vdm-vsg.de

Printed by Books on Demand GmbH, Norderstedt / Germany